THE ETHICS OF GENETIC SCREENING

THE ETHICS OF GENETIC SCREENING

Edited by

RUTH CHADWICK

University of Central Lancashire,
Centre for Professional Ethics,
Preston, Lancashire, U.K.

DARREN SHICKLE

University of Sheffield,
School of Health and Related Research,
Sheffield, U.K.

HENK TEN HAVE

Catholic University of Nijmegen,
School of Medical Sciences,
Nijmegen, The Netherlands

and

URBAN WIESING

Eberhard-Karls University,
Medical Ethics,
Tübingen, Germany

KLUWER ACADEMIC PUBLISHERS
DORDRECHT / BOSTON / LONDON

A C.I.P. Catalogue record for this book is available from the Library of Congress.

ISBN 0-7923-5614-4

Published by Kluwer Academic Publishers,
P.O. Box, 3300 AA Dordrecht, The Netherlands.

Sold and distributed in North, Central and South America
by Kluwer Academic Publishers,
101 Philip Drive, Norwell, MA 02061, U.S.A.

In all other countries, sold and distributed
by Kluwer Academic Publishers,
P.O. Box 322, 3300 AH Dordrecht, The Netherlands.

Printed on acid-free paper

Printed in the Netherlands.

Contents

Contributors

André Boué is a member of the French National Committee of Ethics; Emeritus Professor of Medical Genetics, Paris V University and a Former Director of INSERM Research Unit on Genetics and Prenatal Biology. His research interests include pregnancy wastage, prenatal diagnosis, genetics and ethics in these fields.

Ruth Chadwick is Professor of Moral Philosophy and Head of the Centre for Professional Ethics at the University of Central Lancashire. She has co-ordinated several multi-disciplinary research projects funded by the European Union; Euroscreen 1 - genetic screening: ethical and philosophical perspectives and Biocult: cultural and social objections to biotechnology an analysis of the arguments, with special reference to the views of young people. She is the co-ordinator of the Euroscreen 2 project - Genetic screening and testing: toward Community policy on insurance, commercialisation and promoting public awareness. Her publications include, *Kant: Critical Assessments; Ethics, Reproduction and Genetic Control; the Encyclopedia of Applied Ethics, the Right to Know and the Right not to Know* and a large number of papers in learned journals. She is series editor of the Routledge series on Professional Ethics. She is a Board Member of the International Association of Bioethics, treasurer of the National Committee for Philosophy and a member of the Human Genome Organisation's (HUGO) Ethics Committee.

Angus Clarke studied genetics and clinical medicine as an undergraduate. He worked in general medicine and paediatrics, and then undertook research in clinical genetics. He is now a clinical geneticist with a special interest in neuromuscular disease, ectodermal dysplasia, Rett syndrome and mental handicap. He is also interested in new-born screening programmes and in the social and ethical issues raised by the developments in genetics and biotechnology. He has edited *Genetic Counselling: Practice and Principles* (Routledge 1994) and *Culture, Kinship and Genes* (with Evelyn Parsons, Macmillan 1997), and he has co-authored *Genetics, Society and Clinical Practice* (with Peter Harper, Bios 1997).

Kris Dierickx is from the Centre for Biomedical Ethics and Law in the School of Medicine at the Catholic University of Leuven in Belgium and studied philosophy and moral theology. He is currently preparing a Ph.D. on the ethical aspects of the medical applications of gene technology, with particular reference to genetic testing and screening.

Dolores Dooley is a Senior Lecturer in the Philosophy Department of the National University in Cork, Ireland where she co-ordinates medical ethics and lectures in nursing ethics.

Vladimir Ferak is a Professor in the Department of Molecular Biology at Comenius University in Bratislava, Slovakia.

Gertrud Hauser is Professor of Human Biology at the Faculty of Medicine, University of Vienna. Her research has lain on the border between pathological and normal human variation. Most of it concerns morphological variation in different populations and more recently variation in body composition. She has a long-standing interest in international collaboration, travelling widely and encouraging colleagues, especially in Eastern Europe, to participate in joint research. A former president of the European Anthropological Association, she is a member of the Executive Committee of the International Union of Biological Sciences. This work has brought her face to face with the varying ethical patterns in different countries.

Henk ten Have is Professor of Medical Ethics at the School of Medical Sciences at the Catholic University of Nijmegen, The Netherlands. He is also Visiting Professor at the Centre for Professional Ethics, University of Central Lancashire. His publications include *Geneeskunde tussen droom en drama* (Kok Agora, 1986), *The Growth of Medical Knowledge* (Kluwer, 1990, edited with G. Kimsma and S. Spicker), *Consensus formation in health care ethics* (Kluwer, forthcoming, editor with H. M. Sass) and *Ownership of the human body* (Kluwer, forthcoming, edited with J.V.M. Welie). He is chief editor of Ethiek en Recht in de Gezondheidszorg (Kluwer, 1990 - 1996). He is co-founder and secretary of the European Society for Philosophy of Medicine and Health Care.

Rogeer Hoedemaekers, is a moral theologian and is currently working on his Ph.D. thesis on normative determinants of genetic screening. He is researcher at the Department of Ethics, Philosophy and History of Medicine of the School of Medical Sciences of the Catholic University Nijmegen, The Netherlands, and visiting fellow at the Centre for Professional Ethics of the University of Central Lancashire, Preston, United Kingdom.

Panos Ioannou is a Senior Scientist and Leader of MGUA (Thalassaemia Group) at the Cyprus Institute of Neurology and Genetics. He is currently Principal Research Fellow in the Gene Therapy Group at the Murdoch Institute, Melbourne, Australia. He is Scientific Advisor to the Thalassaemia International Federation; a member of the American Society of Human Genetics, the European Group on Human Gene Therapy, the Human Genome Organization (HUGO) and advisor to the International Atomic Energy Agency in developing a programme on the use of radioisotopes in the prevention of genetic disorders in developing countries. His publications include: P.Trifillis, A. Kyrri, E. Kalogirou, A. Kokkofitou, P. Ioannou, E. Schwartz, S. Surrey, (1993), Analysis of delta globin gene mutations in Greek Cypriots, *Blood*, 82, pp. 1647-1651; P.A. Ioannou, C. Amemiya, P.M. Kroisel, J. Garnes, M. Batzer, P.J. deJong, (1994), pCYPAC-1: A new P1-derived artificial chromosome vector for cloning inserts in the range of 150-300kb in bacterial cells, *Nature Genet.*, 6 pp. 84-89; P.Ioannou, A. Georghiou, E. Bashiardes, M. Kleanthous, G. Christopoulos, T. Foridou, M. Antoniou, C. Paszty, C. Chen, H. Lehrach, M. Angastiniotis, P.J. de Jong, L. T. Middleton, (1995), Sickle cell disease and thalassaemias: new trends in therapy. in Y.Beuzard, B. Lubin, J. Rosa, (eds), Colloque. Gene Therapy approaches in Thalassaemia using P1-derived artificial chromosomes (PACs) and antisense oligonucleotides, *INSERM*, 234, pp. 143- 144; E. Baysal, M. Kleanthous, G. Bozkurt, A. Kyrri, E. Kalogirou, M. Angastiniotis, P. Ioannou, T.H.J.Huisman, (1995), 7-thalassaemia in the population of Cyprus, *Brit. J. Haematol.* 89, pp. 496-499; P. Ioannou, (1996), Gene Therapy for Thalassaemia - which way forward?, *Gene Therapy* 3, pp. 746-747.

Hans-Peter Kröner studied medicine, history, philosophy at Münster University, and now works as a Privatdozent at the Institute of Theory and History of Münster University. He has published on the emigration of Jewish physicians from Nazi-Germany and on the history of eugenics and human genetics.

Veikko Launis is Assistant Professor at the Department of Philosophy, University of Turku, Finland. He is also a member of the Finnish Board for Gene Technology. He is currently working on his PhD thesis dealing with philosophical and ethical aspects of genetics. He is the co-author of *Doing the Decent thing with Genes* (996) and *Genes and Morality* (forthcoming).

Mairi Levitt is a sociologist working in the Centre for Professional Ethics. She is a member of the core group of Euroscreen and was principal investigator on the Biocult project. Her research interests are in attitudes and values in relation to health, education and religion. She has recently published *Nice when they are young: Contemporary Christianity in families and schools* (Avebury, 1996).

Ariadni Mavrou is Assistant Professor of Genetics at Athens University. She studied biology as an undergraduate and obtained her PhD in Genetics in 1984. Her research interests are mainly in the field of cytogenetics and over the years include the study of chromosome abnormalities in recurrent abortions, childhood malignancies, as well as chromosome polymorphisms in the Greek population. She has a special interest in the investigation of Fragile X syndrome using cytogenetic and molecular techniques. Currently her research is focused on the application of molecular cytogenetics in the investigation of certain genetic disorders (Prader Willi and Angelman syndrome, Williams syndrome, Di George syndrome etc.) as well as on the isolation and analysis by in situ hybridisation of fetal cells from maternal circulation.

Caterina Metaxotou is Professor of Genetics at Athens University. She established the first Prenatal Unit for Chromosomal Abnormalities in Greece in 1975. Since 1987 the Unit also performs molecular studies for Neuromuscular disorders, Fragile X syndrome, Prader Willi and Angelman syndromes, etc. She has published over 100 original articles and abstracts in the International Press and has approximately 150 publications in Greek Medical Journals. Her main interest is in providing genetic services to the Greek people and the Genetic Unit provides Genetic counselling to over 3000 families per year. As the only Professor Medical Genetics in Greece she is currently trying to have the field of Medical Genetics recognised as a speciality and to include the subject in the basic curriculum of Greek medical students. She is also involved in the Post Graduate training programmes of Genetics for Doctors and Biologists.

Ingmar Pörn is Professor of Philosophy at the University of Helsinki. In recent years he has published papers on health and disease, emotions and the meaning of life. He is currently working on the conceptual foundations of care and caring.

Judit Sandor is a lawyer, teaching at the Central European University, Budapest Political Science Department. Since 1995 she has been a member of the National Health Science Council of Hungary (Scientific, Research and Ethics Council). In 1996 she gained a Ph.D. in political science and law. Between 1996 and 1997 she was a representative of the Hungarian Patients' Rights Foundation "Szoszolo" and between 1994 and 1997 was course co-director on Human Rights and Medicine at the Inter-University Centre, Dubrovnik, Croatia. During 1993 she was a visiting scholar at the Hastings Center in New York, U.S.A.

Darren Shickle is a Clinical Senior Lecturer in Public Health Medicine at the University of Sheffield. His main research interests are the public health aspects of genetics and medical ethics. He is a member of the core group of Euroscreen, with responsibility for co-ordinating Euroscreen's work on public awareness. He was previously a 1996/7 Harkness Fellow, based at Johns Hopkins University (Baltimore M.D.), Georgetown University, and the Department of Health and Human Services, U.S. Federal Government (Washington D.C.). While in the USA he continued his research interests in public health ethics and he also studied clinical ethics committees and teaching ethics to health professionals. Current research interests include awareness of genetics for public and health professionals, training and education programmes in genetics for the public and health professionals, media coverage of genetics and other public health issues, the impact of genetics on health promotion, and genetic testing in the workplace.

Traute Schroeder Kurth's academic career was at the University of Heidelberg's Institute of Human Genetics. She was appointed Professor in 1974 and Director of the Department of Cytogenetics in 1975. Since her retirement in 1995 she has been Guest Professor at the University of Würzburg. Her research interests include: Fanconi Anemia, Cytogenetics, Prenatal Diagnosis, Genetic Counselling, Ethics.

Urban Wiesing is Chair of Medical Ethics at the University of Tübingen. His publications include *Die In-vitro-Fertilisation - ein umstrittenes Experiment* (Springer, 1991, together with Christina Hölze), *Kunst oder Wissenschaft? Konzeption der Medizin in der deutschen Romantik* (Frommann-Holzboog, 1995), *Zur Verantwortung des Arztes*

(Frommann-Holzboog, 1995). Together with Richard Toellner he edited *Wissen - Handeln - Ethiek. Strukturen ärztlichen Hendelns und ihre ethische Relevanz* (Gustav Fischer Verlag, 1995).

Preface

This collection of essays represents the work produced in the course of a three-year project funded by the Commission of the European Communities under the Biomed I programme, on the ethics of genetic screening, entitled 'Genetic screening: ethical and philosophical perspectives, with special reference to multifactorial diseases'. The short title of the project was Euroscreen, thereafter known as Euroscreen I, in the light of the fact that a second project on genetic screening was subsequently funded.

The project was multinational and multidisciplinary, and had as its objectives to examine the nature and extent of genetic screening programmes in different European countries; to analyse the social policy response to these developments in different countries; and to explore the applicability of normative ethical frameworks to the issues. The project was led by a core group who had oversight of the project and members of which have acted as editors for this volume. Darren Shickle edited the first section; Henk ten Have the second; Ruth Chadwick and Urban Wiesing the third and final part.

The volume opens with an overview of genetic screening and the principles available for addressing developments in the field, with special reference to the Wilson and Jungner principles on screening. The first of the three major sections thereafter includes papers on the state of the art in different countries, together with some analysis of social context and policy. The information contained in these chapters reflects the situation as at November 1997. The second section examines historical and sociological perspectives; the third explores the interplay between normative ethical frameworks and developments in genetics, including analyses of the challenges to our assumptions about applicability of theories and interpretation of concepts.

The Euroscreen II project has taken some of these issues further, with special reference to the implications of genetic information for insurance; the potential impact of commercialisation of genetic screening and testing; and programmes for raising public awareness of the issues.

The editors thank all the participants in the project and gratefully acknowledge the support of the European Commission in funding this project under the Biomed I programme. Special thanks are due to Kerry Wilding, the Euroscreen administrator, for her work in preparing the camera-ready copy.

Ruth Chadwick
Euroscreen co-ordinator

Chapter 1

The Wilson and Jungner principles of screening and genetic testing

DARREN SHICKLE
School of Health and Related Research
University of Sheffield
United Kingdom

1. DEFINITION OF SCREENING

Screening may be defined as a selection procedure for further investigation, applied to a population of asymptomatic individuals, with no personal or family history to suggest that they are at a higher risk of the disease than the rest of the population. The term population when used in the epidemiological sense can be applied to subgroups not necessarily geographically defined, for example according to age, sex or ethnic origin. Thus the definition would also encompass screening for sickle cell disease among a population of Afro-Caribbean origin or Tay-Sachs disease among Ashkenazi Jews.

Screening may be a form of 'secondary prevention', which aims to detect disease in pre-symptomatic individuals in order to provide more effective treatment in the early stages of disease (e.g. neo-natal screening for phenylketonuria), or 'primary prevention' which aims to identify risk factors or carrier states (e.g. carrier screening for cystic fibrosis or predisposition testing for multi-factorial diseases). A screening test should be a quick, cheap and relatively simple procedure to perform, and is used as a sieve for selecting individuals to be offered a 'gold standard' diagnostic test. In view of these time and cost restrictions, the screening procedure will fail to identify people with the disease, trait or risk factor under examination (false negatives). Similarly, some individuals will be incorrectly given a positive screening result (false positives). The implications of these screening errors

1

R. Chadwick et al. (eds.), The Ethics of Genetic Screening, 1–34.

and a more complete discussion of the ethical implications and epidemiology of screening have been covered elsewhere (Shickle and Chadwick, 1994).

2.   GENETIC SCREENING AND GENETIC TESTING

The terms 'genetic screening' and 'genetic testing' are sometimes used synonymously. Strictly speaking, the term genetic screening should only be used for heterozygote or homozygote detection of recessive diseases or for conditions with a high spontaneous mutation rate i.e. where there is no recent family history. This has been described as the 'outwards-in' approach i.e. testing individuals in the community and working inwards to find heterozygotes and then homozygotes (Shickle and Harvey, 1993). Carriers can be detected by an 'inwards-out approach', starting with an index case of a known homozygote or heterozygote and offering testing to family members of individuals known to have an abnormal gene (cascade testing) (Super et al., 1994). Such individuals are at an increased risk of having the gene themselves, for example a 'healthy' brother of a patient with cystic fibrosis is at 66% chance of being a heterozygote himself. In view of their relationship to the known index case then such cascade testing programmes are not covered by the definition of screening. However, many of the methodological and ethical issues that apply to screening are relevant in other forms of genetic testing.

Holloway and Brock (1994) have estimated that cascade screening would identify a high ratio of carriers to people tested (1 in 3 compared with 1 in 25 for general population screening). However, overall fewer than 20% of pregnancies at risk of important single gene disorders can be detected on the basis of a positive family history. Holloway and Brock calculate that cascade screening would only be able to identify 4-13% of all carriers in the population resulting in the detection of 8-24% of all carrier couples. In contrast population cystic fibrosis carrier screening programmes (Mennie et al., 1992; Wald et al., 1993; Livingstone et al., 1994) should be able to detect at least 50% of carrier couples.

The term screening could also be applied to predisposition testing for multifactorial diseases applied to large populations. Individuals tested for dominant or X-linked conditions such as Huntington's disease or Fragile X syndrome will usually have affected family members. Genetic testing for these conditions is usually for diagnostic purposes or part of a case-ascertainment programme. The distinction is an important one, because screening individuals with no previous experience of a disorder or pre-existing expectation that they have a particular gene leads to ethical and psychosocial issues that do not apply in other forms of general genetic testing.

3.   THE WILSON AND JUNGNER PRINCIPLES OF SCREENING

Wilson and Jungner (1968) proposed ten principles of screening in the mid 1960s. These criteria have proved to be useful guidelines for the development of screening programmes and have been widely used:

1. The condition sought should be an important problem.
2. There should be an acceptable treatment for patients with recognised disease.
3. Facilities for diagnosis and treatment should be available.
4. There should be a recognised latent or early symptomatic stage.
5. The natural history of the condition, including its development from latent to declared disease, should be adequately understood.
6. There should be a suitable test or examination.
7. The test or examination should be acceptable to the population.
8. There should be agreed policy on whom to treat as patients.
9. The cost of case-finding (including diagnosis and subsequent treatment of patients) should be economically balanced in relation to the possible expenditure as a whole.
10. Case finding should be a continuous process and not a 'once for all' project.

The Nuffield Council on Bioethics defined genetic screening as:

a search in a population to identify individuals who may have, or be susceptible to, a serious genetic disease, or who, though not at risk themselves, as gene carriers may be at risk of having children with that disease (1993, p. 3).

The Nuffield Council on Bioethics (ibid.) felt that the Wilson and Jungner criteria were not entirely appropriate for genetic screening because the principles were designed for the detection of disease. However as the Nuffield Council recognised, the principles were formulated before pre-natal diagnosis with the associated option of termination of pregnancy came into routine practice. The use of the word 'asymptomatic' within the definition of screening could be interpreted that the individual already has the disease (gene/s) and will become symptomatic in due course. Hence, the Nuffield Council may have been reluctant to apply the Wilson and Jungner criteria to carrier screening for recessively inherited disorders which identifies individuals who are themselves healthy. However, 'asymptomatic' literally means without symptoms, and this is true of carriers. There is certainly no reason for not applying the Wilson and Jungner criteria for homozygote screening for example for phenylketonuria. Thus if principles of screening are considered appropriate for one form of screening, then they should arguably be applied to genetic screening too (whether homozygote or heterozygote).

Wilson and Jungner used the term 'principles' for "ease of description rather than from dogma" (p. 26) It is unlikely that any screening programme will be able to fulfil all of these criteria to everyone's satisfaction. This chapter examines to what extent genetic screening for various disorders meets these principles.

4.   THE CONDITION SOUGHT SHOULD BE AN IMPORTANT PROBLEM

Wilson and Jungner recognised the difficulty in defining 'importance' and included conditions with high prevalence, and those which have very serious consequences for the individual or society as a whole. Thus, the concept of 'important problem' incorporates quantitative and qualitative elements. Screening for a condition which is not 'important' is unlikely to result in net utility, or more utility than other interventions competing for scarce resources. However, it may be worth considering screening for relatively trivial conditions, providing screening is cheap, does not cause anxiety, and there is an effective treatment with few side effects.

The vast majority of genetic disorders are very rare. However, collectively genetic disorders are a quantitatively important public health problem. The birth frequency of a single gene and chromosomal disorders is 2% and between 2-3% of couples have a high risk of a recurring condition in their children (Royal College of Physicians, 1991). In a population survey, Baird et al. (1988) found that 5.5% of the population would develop a genetic or part genetic disorder by the age of 25, and 60% in a lifetime, when common disorders with multiple gene predisposition are included.

Gene or birth frequencies are objective measures i.e. a figure or percentage that can be specified to a number of decimal places (table 1.1). Subjective judgements are only required when deciding when a disorder is sufficiently common to warrant screening. Qualitative definitions of importance and seriousness require more subjectivity and hence scope for disagreement as to when screening is justified.

Rather than recommend the Wilson and Jungner criteria, the Nuffield Council on Bioethics referred to three goals of genetic screening identified by Lappé et al. (1972). It should:

1. contribute to improving the health of persons who suffer from genetic disorders; and/or
2. allow carriers for a given abnormal gene to make informed choices regarding reproduction, and/or
3. move towards alleviating the anxieties of families and communities faced with the prospect of serious genetic disease.

The Nuffield Council on Bioethics believed that further experience of genetic screening could be expected to lead to a more precise definition of principles and goals. However, at the time of writing their report they stated that the prime requirement was that the target disease should be serious. They quoted the Clothier Committee on the Ethics of Gene Therapy which recommended that "the first candidates for gene therapy should be patients who are suffering from a disorder which is life threatening, or causes serious handicap, and for which treatment is at present unavailable or unsatisfactory" (1992, p. 13). The Nuffield Council believed that "such disorders would clearly be classed as serious" (ibid., pp. 17-18). However, they acknowledged that "in the context of genetic screening the definition is likely to be much wider and it is difficult to define precisely what is serious" (ibid., p. 19). They also recognised that the perception of seriousness would vary between societies with and changes in the availability of treatment.

Angus Clarke has proposed that pre-natal testing should be restricted to the most severe disorders, involving either profound retardation, very severe physical handicaps, as in Duchenne muscular dystrophy, or prolonged physical suffering, as in Huntington's disease. Clarke's view falls short of a strict 'right to life position' but he was concerned that about a more explicit agenda of genetic screening:

> While these aims [of clinical genetics] include the offer of reproductive choices, which might result in the "prevention" of some affected individuals, I maintain that the secondary prevention of genetic disorders through the termination of affected or high-risk fetuses should not be an explicit goal. If we include such prevention of genetic disorders amongst our aims, we immediately abandon the non-directive nature of genetic counselling in favour of a genetic public health policy, or eugenics (1991a, p. 999).

Post, Botkin and Whitehouse (1992) suggested that three dimensions of the severity of genetic diseases should be considered: the degree of harm to health if the disease occurs; the patient's age at onset of the disease (early onset considered to be more severe); and the probability that people with the gene will develop the disease (i.e. the penetrance of the gene). Post et al. place more weight on reproductive liberty than does Clarke, but still believed that that some respect should be accorded to presentient fetuses and hence would still prohibit testing that may lead to abortions for 'minor' reasons. According to their view, the pre-natal detection of treatable diseases, late onset diseases, or susceptibilities to diseases does not provide sufficiently weighty reasons to justify abortion. Once again subjective judgements are being used for deciding on what is minor.

One of the problems with severity thresholds is that the judgements of physicians (or policy makers) are being substituted for those of their clients in determining what is a serious enough disease or risk of disease to justify abortion. This may not be a difficulty if such third parties are best placed to judge what is in the individual interest because they are better informed about the disease and the test or are thought more likely to make a rational decision because they are detached from emotional pressures. However, a central goal of pre-natal testing and counselling is to enable couples to obtain information about childbearing risks and use that information to make informed voluntary choices, thereby promoting self-determination in reproductive decision making (Wertz and Fletcher, 1988). Even a severity threshold which would only prohibit testing for very mild disorders would challenge the goal of non-directiveness.

While a third party may be better informed about the condition and the test, they are unlikely to be well informed about the individual's situation, life plans, belief systems etc. A third party would need to second guess the reasons why individuals or couples may want to obtain information from genetic tests and may underestimate the importance of preventing treatable or late onset diseases in some families. Louis Elsas described a case involving a mother of a three year old child with phenylketonuria who had become pregnant again (1990). Despite the fact that she knew that phenylketonuria was treatable after birth since her existing child was already receiving a phenylalanine free diet, the woman requested pre-natal diagnosis with a view to an abortion if the test indicated that the fetus probably had phenylketonuria. The woman justified her request by explaining that care of their affected child involved significant financial and emotional costs. An additional affected children would reduce the emotional and economic resources available to care for each child. Thus, given the choice, they would prefer not to have another child with phenylketonuria.

Carson Strong believed that similar scenarios could be envisioned for late onset disorders such as familial Alzheimer's disease (1997). For example, a couple may request pre-natal testing for Alzheimer's disease if they had seen a relative suffer from the disease for many years. They may wish to avoid further generations being put through the same 'ordeal'. Strong therefore rejected the assertion that the prevention of 'minor' diseases is too trivial a reason to justify abortion, since in some cases at least, the reason may not be at all trivial.

Ideally a classification of severity should not depend on subjective judgements and hence inconsistent 'drawing of lines'. Strong did not believe that all requests for pre-natal testing should be honoured, even though this view would maximise reproductive liberty. He thought that abortion for selection of non-disease characteristics, simply to satisfy the desires of parents, was not ethically justifiable. For example he thought that sex selection merely for parental preference or selection of other non-disease characteristics, such as superior height or above average intelligence were morally trivial reasons for abortion. He recognised that some degree of respect should be accorded to presentient fetuses and that termination of pregnancy for these reasons would perpetuate social attitudes and institutions that result in discrimination and prejudice.

Strong believed that physicians should honour requests for pre-natal tests for diseases, including relatively minor diseases and susceptibilities to diseases, but not requests pertaining to non-disease characteristics. While, as he points out, it may be easier to distinguish disease from non-disease than to distinguish diseases that are 'too minor' from those that are 'not too minor', the 'distinguishing' process still requires subjectivity. He acknowledged that pre-natal testing for morbid obesity might be motivated by a desire to avoid health problems related to obesity or, alternatively, to avoid an unattractive physique. In other cases, for example low intelligence or short stature, it is not always clear whether they are pathological or merely variations within normal limits. Despite making this concession, Strong believed that it was usually possible to make a clear distinction between conditions that are diseases such as phenylketonuria or cystic fibrosis and those that are not such as gender or height, based on patho-physiology. However, the fact that a particular condition is of interest to a pathologist or physiologist is not an adequate definition of disease. The on going debate on the status of mental illness as a disease, is in part due to an inability to point to patho-physiological changes (Fulford, 1989).

Strong suggested that requests for tests for minor diseases could be "unreasonable because they reflect a lack of knowledge by patients concerning such factors as the risks associated with testing, the probability that the fetus is affected, and the severity of the disease in question" (ibid., p.146). He argued that some degree of directiveness was defensible because it reflected what he thought was a reasonable viewpoint concerning the purposes of reproductive genetic counselling i.e. to focus on helping patients deal with diseases even if this includes minor diseases or susceptibilities to diseases. Providing more information would be in keeping with assuring informed decision making. However, if additional 'counselling' fails to persuade the client, what then? By making this qualification, Strong is facing the same tension as Clarke and Post et al., between avoiding termination for trivial reasons and reproductive autonomy. The only difference is that Strong is nearer to the 'reproductive liberty' end of the continuum.

5.   THERE SHOULD BE AN ACCEPTABLE TREATMENT FOR PATIENTS WITH RECOGNISED DISEASE

Wilson and Jungner felt that "of all the criteria that a screening test should fulfil, the ability to treat the condition adequately, when discovered, is perhaps the most important" (ibid.).

When we think of medical treatments, we normally think of a doctor prescribing a drug or performing surgery or some other procedure, the subject of the treatment being a patient.

Phenylketonuria is perhaps the only relatively common genetic disease that can be effectively treated. Although there is no firm evidence, it is generally agreed that a restricted phenylalanine intake should be continued as long as family and social conditions allow. The high levels of phenylalanine are only clinically significant during the period of neurological development. Careful control of diet should certainly continue at least until 10 years of age but after that time restrictions can be relaxed. The high blood concentrations of phenylalanine would therefore be of little consequence to adults, except during pregnancy when phenylalanine can cross the placenta to affect neurological development of the fetus. Women with phenylketonuria should therefore go back on the restriction diet prior to attempting conception until termination of breast feeding.

The Nuffield Council on Bioethics were correct to recognise that the perception of severity of a particular disease may change over time as new treatments become available. For example, birth cohort analysis for cystic fibrosis shows that the survival of more recent birth cohorts continues to improve - probably due to a combination of earlier detection and improved treatment - so that the median survival of children born with cystic fibrosis now is likely to be substantially better, perhaps as high as 40 years (Elborn et al., 1991). As an indicator of functional capacity, a recent study of UK cystic fibrosis sufferers over 16 years old, showed that 54% were in paid employment (Walters et al., 1993).

Hitherto, there was little that could be offered patients with genetic disorders. Even the dietary control of phenylketonuria could not be a considered a cure. It is hoped that gene therapy may provide the opportunity for more effective treatment or even a cure. Gene therapy is a therapeutic procedure in which genes are intentionally inserted into cells using vectors. The function of the vector is to protect the therapeutic genes and to transport them safely into the nucleus of the target cells, where (hopefully) the DNA is inserted into the correct position within the genome to be expressed normally. Gene therapy can be performed in vivo whereby genes are delivered directly to target cells in the body, or ex vivo in which the target cells are genetically modified outside the body and then re-implanted.

At present gene therapy is only able to add genes that were absent or non-functional, and cannot delete or remove defective genes whose products are contributing to the disease pathology. Current gene delivery systems are also inefficient in introducing the gene into sufficient target cells to achieve a therapeutic benefit. Because of these limitations, research efforts have concentrated on a few genetic deficiency diseases such as cystic fibrosis, haemophilia, and severe combined immune deficiency in which the target cells are relatively accessible.

The first human gene transfer experiment was reported in 1990, when a patient with malignant melanoma received genetically modified autologous T cells (Rosenberg et al.). Since then about 30 gene therapy companies have been launched, three major gene therapy journals have been established, more than 200 human gene therapy protocols have been approved, and over 2000 patients have received gene therapy (Russell, 1997). The diseases most often treated with gene therapy are cancer (68%), AIDS (18%), and cystic fibrosis (8%). Gene therapy can be used in different ways to destroy unwanted cells such as cancer cells or HIV infected cells. One strategy is to introduce a drug susceptibility gene ('suicide' gene) into target cells so that they are selectively killed when the appropriate drug is subsequently administered.

The most successful use of gene therapy research so far was the treatment of severe combined immune deficiency secondary to adenosine deaminase deficiency by reinfusing genetically corrected autologous T cells. One of the two affected children treated has had a full and sustained recovery of a whole range of immunological parameters and is now able to lead a normal life. However, despite the early optimism only a handful of patients with rare conditions have benefited from gene therapy.

However, it is important not to be too restrictive in our thinking about what represents an 'adequate' treatment. For example, with some cases of pre-natal diagnosis, couples may benefit from genetic counselling such that they may 'prepare themselves' for the birth of an affected child (Jones et al., 1988). In this circumstance, the counselling could be considered to be a form of treatment. An analogous question is whether termination of pregnancy can be considered a treatment, and if so for whom. However, the debate on that question has been explored elsewhere (Clarke, 1993).

The main purpose of testing for fragile X (Murray et al., 1997) and Duchenne muscular dystrophy (Bradley et al., 1993) is to provide information and counselling for other family members to allow them to make more informed choices about their own reproduction. A child with Duchenne muscular dystrophy usually develops and grows normally until the age of 18 months of age and by the time that a formal diagnosis is made, his parents may have given birth to further affected children. The individuals detected during neo-natal screening have little or nothing to gain by such testing, other than being saved the prolonged series of investigations when they start to become symptomatic.

Information obtained from genetic testing may be used within the provision of health promotion advice. The risks of smoking are well known in the general population, but despite this, a quarter of adults are regular smokers. This contradiction may indicate that smokers are unable or unwilling to convert population statistics into personal risk. Epidemiological data can be more closely tailored to an individual if stratified by age, sex and cigarette consumption history. However, genetics provides a potentially more powerful tool for individualising risk. For example, genes coding for enzymes involved in the metabolism of carcinogens in tobacco may indicate predisposition for lung cancer. In the future it will be possible to quantify an individual's predisposition for a range of multifactorial diseases.

It is unclear as to the impact that this will have on health promotion. For example, it might be expected that an individual with a demonstrated increased genetic predisposition to lung cancer might be more likely to attempt to reduce other risk factors, where this is possible, such as cigarette smoking. However, they may become fatalistic, and conclude that they may as well continue to smoke, if they are 'going to get lung cancer anyway' (Lerman et al., 1997). In addition, current health promotion advice on smoking can describe the benefits of reduced risk for a number of diseases. In the future, the 'bad news' of increased risk of one disease may be balanced by reduced genetic risk of others. Individuals may find it very difficult to assimilate all the various risk data and prefer to place more emphasis on information about reduced risks when making lifestyle choices.

6.   FACILITIES FOR DIAGNOSIS AND TREATMENT SHOULD BE AVAILABLE

The importance of availability of facilities for diagnosis and treatment is self-evident. Informing a patient that they have a treatable disease, but withholding any intervention is likely to cause more harm than not performing screening. Until gene therapy becomes more effective, the main intervention that can be offered post screening is genetic counselling, providing information that may increase reproductive autonomy. However, one of the themes that appears in many of the country reports in section A is the poor availability of counselling skills. Any expansion in genetic testing will need to be accompanied by an expansion in the number of health professionals with the necessary counselling skills.

7.   THERE SHOULD BE A RECOGNISED LATENT OR EARLY SYMPTOMATIC STAGE

Last defined latent period as the "delay between exposure to a disease-causing agent and the appearance of manifestations of the disease." (1988, p. 72). As an example Last quotes an average latent period of five years between exposure to ionising radiation and the development of leukaemia, and more than 20 years before development of certain other malignant conditions.

As a Norwegian report has pointed out:

New forms of genetic testing distinguish themselves from other types of medical testing by giving the same result independent of age, independent of medical condition, independent of biological material and independent of the amount of the tested material. (Ministry of Health and Social Affairs, 1993, p. 52)

The Norwegian report suggested that the result would be the same regardless of when in life the test is performed "because human genes remain unchanged from conception to death" (ibid.). On this simplistic account then it would be very easy to identify a latent stage, since it would begin at conception. Last also defined latency as "the period from disease initiation to disease detection" (ibid.). Since it is increasingly recognised that there is a genetic component to most if not all disease, then life would be one continuous disease process. However, the Norwegian definition ignores mutations that occur in the nuclear and mitochondrial genome throughout life, which can result in disease. Thus, a negative test result at an early stage of life should not provide complete reassurance.

8.   THE NATURAL HISTORY OF THE CONDITION, INCLUDING ITS DEVELOPMENT FROM LATENT TO DECLARED DISEASE, SHOULD BE ADEQUATELY UNDERSTOOD

The requirement for an understanding of a condition's natural history and recognition of a latent stage, is important in the development of techniques for diagnosis and treatment, but in themselves is not essential. Screening for HIV is performed, yet the natural history of acquired immunodeficiency syndrome (AIDS) is still not 'adequately understood'.

The Human Genome Project was established with the aim of sequencing the entire human genome by 2005. However it is only the start of the project; it is projected that it will take the remainder of the twenty-first century to understand what each gene does, how they are controlled, and how the genes interact with one another. This further work is likely to alter the way in which we classify disease. One of the challenges to clinical medicine is that patients with the same diagnosis have very different disease manifestations and responses to therapy. This may be because different genotypes may produce similar phenotypes, such that historically they have been classified as the same disease. Diseases classified by genotype may have narrower clinical presentations and would respond better to more targeted and specific therapies.

All cystic fibrosis is due to mutation in the chromosome 7 CFTR locus and not in any other gene. Consequently, the variation in clinical severity must arise from the type of mutations at that locus, the effects of other modifying genes or non-genetic factors. Kerem et al. examined the relationship between genotype and phenotype (Kerem et al., 1990). Patients homozygous for the common ΔF508 mutation are usually more severely affected that those heterozygous for ΔF508, or who have two other alleles. For example 99% of (ΔF508 / ΔF508) patients had pancreatic insufficiency compared to 40% of (ΔF508 / other allele) patients and 36% of patients who had no ΔF508 mutations. Similarly homozygous ΔF508 patients were more likely to be diagnosed at a younger age (mean of 1.8 years, compared with

4.4 and 8.4 years for ΔF508 heterozygotes and patients without a ΔF508 allele, respectively).

The Norwegian report suggested that the result of a genetic test will, in most cases, be the same whether the individual is already ill, or whether the disease will give symptoms in fifty years time. However, it goes on to say that "the result will also be the same if the illness is mild or if it is very serious, or even if the disease does not manifest at all." (ibid.). In this respect, the concept of latency in genetic disease is complicated by the very fact that even though an 'abnormal' gene has been identified it does not guarantee that a disease will develop. For some diseases, the identification of a particular gene is only an indication that an individual is at increased risk of developing a particular disease, rather than the identification of an asymptomatic individual during the latent period.

9.   THERE SHOULD BE A SUITABLE TEST OR EXAMINATION

There are four main opportunities in the life-cycle for screening, each with advantages and disadvantages (table 1.2).

Historically, population genetic screening was only possible by testing for phenotypic manifestations. For example, patients with phenylketonuria were identified by measuring phenylalanine levels, cystic fibrosis could be diagnosed by measuring sodium or chloride in sweat, or sickle cell disease could be identified by looking for the characteristic deformed blood cells on a blood film. For some genetic disorders e.g. sickle cell disease, thalassaemia, Tay-Sachs disease, carriers could also be detected by haematological or biochemical techniques.

It is not always possible to make an absolute distinction between those individuals where the disease is present or absent even with a 'gold standard'. For some conditions there will be a continuous distribution of variables, such that at one extreme, individuals could be considered 'diseased', while those at the other are 'healthy', e.g. hearing or visual acuity. For other biological variables e.g. blood pressure or haemoglobin, the extremes at both ends of the distribution would be 'unhealthy', while those in the middle would be 'normal'.

For the purposes of screening there should be a threshold to trigger further investigations or treatment. The position of this cut-off in the distribution should be based on the associated risk of morbidity or mortality that warrants further intervention. The choice of threshold may therefore be arbitrary or depend on the resources available. For example, a blood cholesterol of 6.5 mmol/l or above, is considered to be associated with a significantly high risk of coronary heart disease (Standing Medical Advisory Committee, 1990). However, between 25 and 36% of the UK population have a blood cholesterol above this threshold, and it would be unrealistic to treat this number of people.

In the early 1970s it was shown that raised amniotic fluid α-fetoprotein (AFP) was associated with open neural tube defects. It was later demonstrated that raised AFP in maternal blood was also indicative of neural tube defects (NTDs). Maternal serum AFP screening for NTDs became commonplace. However, it was subsequently noted that pregnant women with low serum AFP measurements were at increased risk of having a baby with Down's syndrome. Other serum markers were also found to be associated with Down's syndrome: e.g. raised human chorionic gonadotrophin (hCG) and lowered unconjugated oestriol (uE$_3$). A combination of such markers together with the woman's age and fetal gestational age (determined by ultrasound scan) can be used as the basis of a Down's syndrome screening programme, with the diagnosis confirmed by amniocentesis (Wald, et al., 1998). However, some women pregnant with a fetus affected by Down's syndrome had higher serum AFP results than some women whose fetuses were not affected with Down syndrome. An arbitrary cut-off must therefore be chosen: if it is set at a high concentration of AFP there will be many false positives (i.e. women found to have an unaffected fetus following amniocentesis), if it is too low there will be more false negatives (i.e. undiagnosed babies born with Down's syndrome).

The performance of a screening test is measured by calculating sensitivity and specificity (table 1.3)

A screening test with high specificity would only tend to detect people with the disease i.e. relatively few false positives. Raising the threshold would raise specificity, and hence target resources at those with highest risk of morbidity and potentially most to gain from treatment. In contrast, a test with high sensitivity would tend to maximise identification of diseased people in the screened population i.e. relatively few false negatives, but there would also be unnecessary investigations or treatment for others. Thus an increase in sensitivity will be at the expense of specificity and vice versa.

The preference for specificity or sensitivity should depend on the consequences of making the diagnosis or not. High specificity should be desirable if:

1.   the diagnosis is associated with anxiety or stigma;
2.   further investigations are time consuming, painful and/or expensive;
3.   cases are likely to be detected by other means before it is 'too late' for effective treatment;

4.   treatment, especially if painful or expensive, is to be offered without further investigations.

In contrast high sensitivity would be desirable if:

1.   adverse consequences of missed diagnosis for the individual e.g. late treatment may be significantly worse than early;

2.   adverse consequences of missed diagnosis for society e.g. with a serious communicable disease;

3.   diagnosis is to be confirmed by other investigations, so that the period of anxiety is short, or the correct diagnosis is given before treatment is started.

Such trade-offs between sensitivity and specificity will be dependent on methodological and economic factors, however, they are also intrinsically moral. For example, one outcome of Down's syndrome screening is that fewer children with Down's syndrome are born, because many parents who are informed by the screening the process that their child is affected choose to terminate the pregnancy. However, as part of the diagnostic process some fetuses (the majority of whom will be healthy) will be lost through spontaneous miscarriage, which is a recognised complication of amniocentesis (0.5-1%).

When a screening programme is evaluated, the benefits of being a true positive tend to be considered, and sometimes the reassurance associated with a negative result. However, associated costs tend to be ignored. If costs outweigh benefits then either the balance between sensitivity and specificity should be altered or else the screening test should not be used at all. Screening programmes for Down's syndrome depend on the moral conclusion that the loss of healthy fetuses in order to detect those affected by Down's syndrome is a price worth paying.

Examples of costs and benefits of screening associated with each of the four permutations of true/false positive/negative were identified by Shickle and Chadwick (1994) (table 1.4).

In more recent years, developments in genetic technology have meant that it is possible to screen genotype directly. In circumstances where all individuals affected by the disease have the same genotype i.e. where there is only one disease causing mutation that occurs in one gene, sensitivity and specificity are very high because either the abnormal gene is present or it is not. However this is not always the case. There may be many mutations seen within the gene, each of which requires a separate gene probe, or there may be more than one gene within the genome where malfunction can lead to the disease.

The cystic fibrosis gene was cloned in 1989 (Rommens et al.). By 1992 over 150 different mutant cystic fibrosis alleles had been described (Mennie et al.), increasing to over 400 by 1994 (The Cystic Fibrosis Genetic Analysis

Consortium). The large number of mutations presents a problem when devising a suitable test for mass screening for cystic fibrosis. Different primers are required for the PCR for each mutation to be identified. In the future DNA chips will permit screening for thousands of mutations simultaneously. However, at present, the commercially available kits are only able to detect the more common mutations (Shickle and Harvey, 1996).

The CF Genetic Analysis Consortium published data in 1994 of the results of cystic fibrosis gene mutations testing worldwide by its member Centres. A total of 43,849 cystic fibrosis chromosomes were tested for up to 24 of the most common mutations (not all Centres tested for all mutations). The data was not truly representative of the world gene pool, since almost a half of patients tested were from Northern Europe. In total, about three quarters of the chromosomes tested had mutations identified. Over eighty percent were detected in Northern Europe and Australasia, while only about a half of chromosomes tested in Central and South America had mutations identified. The most single common mutation world wide is ΔF508 (66%), but again this statistic is biased by a much higher rate of measurement in North Europe. The implications of this will be discussed in more detail in Section A.

The problem of testing for many mutations simultaneously is even more complicated when there are many loci (positions in the genome) where mutations can cause a particular disease. Five to ten percent of breast cancer is attributable to the autosomal inheritance of a high risk susceptibility gene. There are a number of known inherited cancer syndromes that confer a higher risk of breast cancer. Recently, the BRCA1 gene, which is responsible for 45% of hereditary early onset breast cancer and for the majority of co-inheritance of breast and ovarian cancer, has been cloned. Another gene that confers an increased risk of breast cancer is the BRCA2 gene, which maps to the long arm of chromosome 13. Mutations in BRCA2 account for approximately 40% of hereditary early onset breast cancer. In addition, at least 7% of breast cancer may occur in women who are heterozygous for mutations in a gene for ataxia-telangiectasia, an autosomal recessive chromosome instability syndrome.

BRCA1 is a putative tumour suppressor gene located on chromosome 17q21. It spans 100kb of genomic DNA and encodes a protein of 200kD consisting of 1863 amino acids. Sixty-three distinct germline mutations of BRCA1 have now been identified in more than 100 patients with breast and/or ovarian cancer. These mutations are distributed across the entire coding region of the BRCA1 gene. At present, it is therefore not feasible to provide meaningful information about breast cancer risk unless a woman has an affected family member who can provide genetic material.

10.   THE TEST OR EXAMINATION SHOULD BE ACCEPTABLE TO THE POPULATION

The acceptability of a screening test to any particular individual will depend on how they perceive the size and nature of the costs and benefits associated with the test. The way in which individuals perceive risk is complex: some are 'risk lovers' while others are 'risk averse' or 'risk neutral' (O'Brien, 1986). Counselling should, as far as possible, facilitate the autonomous decisions of the individual, e.g. allowing informed consent to be obtained for screening. It is unlikely however, that any individual could absorb all the necessary information during the limited time allocated for pre-test counselling (if any at all), especially if this is during a time of anxiety such as during pregnancy. At the extreme pre-test counselling may amount little more than asking: 'Do you want to be tested to see whether you (or your child) has a disease that can cause death (or severe disability), for which we **may** be able to offer you some intervention'. Put in such terms it is understandable that an individual is likely to place considerable weight on the advantages of being a true positive or true negative. A test associated with high costs (both tangible and intangible) may be acceptable to an individual, if the condition is perceived as severe and there is an offer of available treatment (even though the individual may have limited capacity to evaluate it's effectiveness).

The harm consequent from a screening test for any individual will usually be trivial, in comparison with the potential harm from not offering screening and saving a life by providing effective treatment. For example, most women will endure the indignity and embarrassment associated with taking a cervical cytology smear, if it means that carcinoma of the cervix may be prevented. The acceptability of genetic screening benefits from the fact that physical discomfort associated with obtaining test material is limited, since samples of DNA may be obtained from mouthwash or buckle smears.

An individual may choose screening even though there is a high risk of relatively trivial harm and only a small chance that they will benefit. Using utilitarian principles, the net utility or disutility for society arising from an intervention is equivalent to the sum of the change in utility to its individual members. However, the sum of trivial disutility for the many may exceed the large utility gain for the few. The contrast in conclusion between suitability (for a population) and acceptability (to the individual) can be problematic when devising public policy if both perspectives are considered. For example a National Institutes of Health Consensus Panel recognised that there was insufficient evidence of effectiveness to recommend mammography screening for 40-50 year olds, but nevertheless recommended that women in this age range should be permitted to have mammography if they requested screening (1997).

Uptake of screening is the most direct measure of test acceptability. Shortly after the identification of the gene responsible for cystic fibrosis, Williamson et al. conducted a survey of public attitudes to carrier screening (1989). They found that only a minority had clear pre-existing knowledge of cystic fibrosis and its genetic nature. However, over 80 per cent of those questioned expressed interest in knowing their carrier status. They acknowledged that uptake can only be assessed when a service is in place, however they concluded that their results strongly suggest that there will be interest in community-wide testing for cystic fibrosis carrier status when such a test became available.

The uptake rate of cystic fibrosis carrier screening has varied widely from one pilot programme to another. Even within studies, very large variations in uptake have been observed from one screening location to another. Hartley et al. conducted screening within eight different general practices: uptake varied from 11 to 99% (1997). A large degree of this variation was explained by the way in which the offer of screening was made. For example Watson et al. offered screening in two general practices and four family planning clinics (1991). When screening was offered opportunistically the uptake was 66% in general practice and 87% in family planning clinics. However, only ten per cent of those offered a screening appointment by letter took up the invitation. Another cystic fibrosis carrier screening pilot study performed in a primary care setting also found that the uptake rate varied with mode of invitation (1997).

Bekker et al. specifically examined different ways of offering screening within one location (1993). Only six percent of patients who received a letter from their GP inviting them to contact the practice to make an appointment did so. Other patients were handed a leaflet by the receptionist when visiting the practice for some other reason offering them the opportunity to be tested immediately. Seventeen percent of these patients offered passive opportunistic screening were tested. However, if patients in the waiting area were approached by a member of the research team, told about the test personally, and invited to undergo testing at that time, the uptake rate was significantly higher (70%). Bekker et al. suggested that their research had implications for the degree of consent obtained for screening and the level of enthusiasm for screening:

Our study does not allow us to assess the degree to which those who respond to a personal approach are motivated by a desire for carrier status information, willingness to accept advice from a health professional provided that it does not entail too much inconvenience, or unwillingness to refuse a polite request from a health professional. The finding that most patients who accept the offer on one day but fail to return on another day for the test supports one or both of the second interpretations. (p. 1586)

Bekker et al. concluded that the public unwillingness to make any more than minimal effort to be tested indicated a "supply push rather than a demand pull". Poor uptake levels may reflect apathy or insufficient knowledge about the condition for which testing is being offered. However, even among families affected by autosomal dominant conditions, where they will have first hand experience of the disease and its impact, uptake of genetic testing has been surprisingly low.

Meissen and Bercheck surveyed individuals at a 50% risk of Huntington's disease shortly after a marker for the gene that causes Huntington's disease had been discovered. Their objective was to assess likely uptake of testing if a predictive testing programme was offered. A majority (65%) of subjects favoured being tested. However, an actual offer of testing is very different from a hypothetical inquiry. Tyler et al. reviewed the experience of genetic centres in the UK with pre-symptomatic testing for Huntington's disease between 1987 and 1990 (testing in the UK was first performed in 1987) (Tyler et al., 1992). They commented that demand for testing was "much less than expected as surveys carried out before linkage was possible suggested that between 56% and 66% of people at risk would make use of a predictive test if it were available" (p 1594). In practice only 248 tests were performed in the first years of test availability, which is a very small proportion of the estimated 10,000 people at a 50% risk of the disease. While low uptake does not mean that the test is unacceptable to the population at risk, it does demonstrate that many members of the target group prefer hope (that they are unaffected) versus certainty (whether they carry the gene of not). The balancing of 'a right to know and a right not to know' has been discussed elsewhere (Chadwick, et al., 1997).

11.   THERE SHOULD BE AGREED POLICY ON WHOM TO TREAT AS PATIENTS

Although Fragile X syndrome is X linked, the underlying genetic mechanism is not straightforward. The gene promoter region contains a CCG repeat island with a repeat length, normally in the 20-40 range, that is passed down from generation to generation. The repeat length may slowly increase over generations and become progressively more unstable, probably because of 'slippage' (inaccurate duplication that occurs in stretches of identical repeats). An elongation of 54 or so repeats can suddenly expand to more than 200 in one generation. This large change, or full mutation, switches off the promoter and stops gene production. Both sexes may have smaller expansions or pre-mutations of 54-200 repeats. These people are normal (but the women have an earlier menopause). They are called 'normal transmitting' males or females. Expansion or amplification into the full range of mutation occur only when premutations are transmitted through females. The variability in length of CCG repeat length means that the cut off point for treating individuals as 'patients' is arbitrary.

12. THE COST OF CASE-FINDING (INCLUDING DIAGNOSIS AND SUBSEQUENT TREATMENT OF PATIENTS) SHOULD BE ECONOMICALLY BALANCED IN RELATION TO THE POSSIBLE EXPENDITURE AS A WHOLE

The balancing of economic costs resulting from case-finding with those due to possible expenditure as a whole, is compatible with a modified utilitarian approach and with the principle of justice.

Phenylketonuria was the subject of one of the earliest attempts within health economics to perform a cost-benefit analysis (Bush et al., 1973).

The prevention of severe mental retardation, apart from the human benefit, releases the community from providing expensive long-term medical and social supervision for an individual who is instead able to lead a normal and independent life (Kromrower, et al., 1984). The Department of Health Register of Cost-effectiveness Studies have converted Berg et al's original 1970 US $ cost estimates to give an incremental cost per QALY at 1991 UK prices of £288.4 per infant (discounted at 4%) (Department of Health Economics and Operational Research Division, 1984, p. H1). This figure compares favourably with other data on cost-per-QALY. For example, in 1991, the cost per QALY from pacemaker implantation for heart block was £700; hip replacement £750; GP control of total serum cholesterol £1,700; kidney transplantation £3000; breast cancer screening £3500; heart transplantation £5000; and hospital haemodialysis £14,000 (Mooney, 1992).

A number of cost-benefit and cost-effectiveness studies have been performed of genetics services (Dhondt, et al., 1991; Chapple, et al., 1987).

While these studies have usually been unable to measure intangible costs and benefits, they consistently show that genetic screening would save money for the NHS. For example, the life time cost to the NHS of caring for one patient with Down's syndrome has been estimated to be double the cost of examining 100 amniotic fluids and terminating one fetus (Gill, et al., 1987).While the use of QALYs has been defended from an ethical standpoint (Edgar et al., 1998), any use of economic data within the context of genetics raises concerns of eugenics (Kelves, 1985).

Clarke expressed his concerns that pre-natal diagnosis and subsequent termination of affected fetuses would be used "to justify ... receipt of public money" (Clarke, 1991b, p.1145), rather than "the prevention of future suffering, when this is sought by the families concerned" (ibid., p. 1145). He is supportive of audit of process to improve quality of case management and audit of outcome to ensure that clients are satisfied with the service offered. However, he believes that outcome audit poses ethical difficulties:

An unacceptable measure of outcome, for example, would be utilisation of population-based data, (e.g., numbers of terminations for a specified condition or a trend in birth incidence). If a department's work is to be measured in such terms, there will be subtle - and possibly less than subtle - pressure upon clinicians to maximise the rate of terminations of pregnancy for 'costly' disorders: a completely unacceptable outcome which we must strive to prevent (ibid., p. 1146)

There were some measures of outcome acceptable to Clarke, for example calculations of the normal male fetuses 'spared' from termination by the existence of molecular genetic services which demonstrated that they are not affected by Duchenne muscular dystrophy. He also suggested estimating the number of healthy fetuses conceived by couples at risk of having an affected child, who may not have risked a pregnancy without the reassurance of the availability of pre-natal diagnosis.

It is possible to apply economic analysis to genetics evaluation, without depending on termination of pregnancy as a 'major cost-saver'. For example, Phin (1990) proposed cost/case detected or the number of amniocenteses/case detected rather than cost/case aborted or unit of disability averted.

Modell and Kuliev (1991) have suggested an approach to economic analysis using informed choice as the main benefit of genetic screening. Although disease prevention was used as a desired goal, it was excluded from their analysis of costs and benefits. Their modified economic analysis still demonstrated the net benefit of genetic screening. Thus it is possible to advocate purchasing of genetics services in terms of health gain through 'non-directive' counselling, with the added advantage that one option of 'informed choice' is termination of pregnancy which will result in a saving in health care spending.

13.   CASE FINDING SHOULD BE A CONTINUOUS PROCESS AND NOT A 'ONCE FOR ALL' PROJECT

If screening is available for one individual, then justice would expect screening to be offered for another individual who has a similar risk of having the disease and potential to benefit from treatment. If a screening programme is considered to be desirable, why then should future patients not benefit from screening? Wilson and Jungner argued that screening should be a continuous process, since the 'start-up' costs associated with screening (providing accommodation, purchasing equipment, training staff etc.) are usually large in comparison with the marginal costs of each test performed. While these points are true, screening for prevalent (existing) cases require different considerations than for incident (new) cases. Thus, the case yield and the consequent utility may be high in the first wave of screening if the natural history of a condition is slow and hence there may be a large number of undetected cases in a population. However, the number of new cases that develop during a screening cycle may be small and hence the consequent utility for these few cases may not outweigh disutility or opportunity costs.

The initial testing round of a genetic screening programme should identify all individuals with the gene(s) if interest among those tested. It take many years to screen all the individuals in a community who want to be tested. However, once this process is complete, the screening programme would only need to be available to test newborns, or children as they reach the age when they are able to give consent, or people who migrate into the community. The number of tests performed and hence the number of cases detected in this second phase will mean that different conclusions about cost effectiveness may be drawn and either the screening programme needs to be modified to adapt to the new requirements, or it may be abandoned completely.

14.　　OTHER GUIDELINES FOR EVALUATING SCREENING

While the Wilson and Jungner criteria are the best known and most frequent used principles, other principles have been proposed. For example Cuckle and Wald proposed eight principles (1984) (table 1.5).

Cochrane and Holland suggested seven criteria for assessment and evaluation of a screening test (1971):

1.　Simplicity: a test should be simple to perform, easy to interpret, and where possible, capable of use by paramedical and other personnel. With increasingly complex technology certain screening tests, particularly for example in the ante-natal and neo-natal periods, can only be performed by doctors.
2.　Acceptability: since participation in screening is voluntary, a test must be acceptable to those undergoing it.
3.　Accuracy: a test must give a true measurement of the condition or symptom under investigation.
4.　Cost: the expense of the test must be considered in relation to the benefits of early detection of the disease.
5.　Precision or repeatability: the test should give consistent results in repeated trials.
6.　Sensitivity: the test should be capable of giving a positive finding when the person being screened has the disease being sought.
7.　Specificity: the test should be capable of giving a negative finding when the person being screened does not have the disease being sought.

15.　　CONCLUSION

Genetic testing is already widespread in Europe. Virtually all neonates are screened for phenylketonuria and congenital hypothyroidism. Most pregnant women are offered screened for Down's syndrome and neural tube defects either using an age cut-off or by serum screening. Congenital anomaly scans using ultrasound are increasingly common to screen for congenital malformations. Haemoglobinopathy screening is routine in some southern European countries and in many large cities throughout the continent which have large ethnic minority populations from parts of the world where haemoglobinopathies are common.

This workload will expand further as carrier screening for cystic fibrosis is introduced. In northern European Caucasians the incidence of cystic fibrosis varies from 1 in 2000 to 1 in 3000 births (i.e. about 1 in 25 of the population are carriers. The gene frequency of cystic fibrosis in Eastern and Southern Europe is slightly lower, which means that there is a weaker case for cystic fibrosis screening.

As we understand the genetics of multifactorial diseases there will be an exponential increase in genetic testing. For example 1 in 12 women in the UK develop breast cancer at some point during their lifetime: in 1989 there were 27,768 women were registered as being diagnosed with malignant neoplasm of the breast and in 1995 there were 12,543 deaths from breast cancer. Section A of this book will provide an overview of genetic screening in Europe, and highlight issues in existing and proposed genetic screening programmes in specific European Countries: Austria, Belgium, Cyprus, Finland, France, Germany, Greece, Ireland, Netherlands, and Slovakia.

It is unlikely that many of the existing genetic screening programmes in Europe have been evaluated against the Wilson and Jungner criteria or it's equivalents. Holland and Stewart (1990) grouped the various screening principles into four categories: condition, diagnosis, treatment, cost. In summary, therefore, is screening for genetic disorders supported by the various evaluation criteria?

15.1   Condition

While genetic disorders are collectively common, many are too rare to support individual population screening programmes. Certainly carrier screening for cystic fibrosis or genetic cancer susceptibility are conditions that are attracting attention for possible population screening in the near future. It may be possible to combine testing for rarer conditions, for example as in neo-natal screening using Guthrie blood spots. For rarer or autosomal dominant conditions such as Huntington's disease, cascade testing may be more appropriate.

15.2   Diagnosis

Obtaining samples to extract DNA is cheap and straight forward (e.g. by mouth wash, or buccal smears). Testing for diseases involving many mutations or gene loci represents a major hurdle for mass screening (e.g. for cystic fibrosis or breast cancer). Public acceptability as indicated by uptake, suggests that there may be limited demand for genetic testing, unless the disease is sufficiently severe. Defining the severity of the disease that warrants genetic testing remains contentious.

15.3   Treatment

Treatment options for most genetic disorders remain limited. While gene therapy is promising, the research is yet to be produce significant commercial developments. The main outcome of genetic screening therefore remains the provision of information that assists in reproductive decision making. Abortion remains a highly contentious subject. If the provision of information both pre- and post-test means that the quantity and quality of counselling support is crucial.

15.4   Cost

In view of this potential demand for genetic screening, it will be important to prioritise scarce health care resources (both in terms of skills and money). The history of genetics and the ethical, legal and social implications of genetic testing also mean that genetic screening programmes require particular attention and these subjects are discussed in more detail in sections B and C.

The Wilson and Jungner criteria therefore represent an important aid in the assessment of genetic screening programmes to ensure that they are methodologically, economically and ethically sound.

REFERENCES

Baird, P.A., Anderson, T.W., Newcombe, H.B., and Lowry, R.B., 1988. 'Genetic disorders in children and young adults: a population study'. *American Journal of Human Genetics*, 42, pp. 677-93.

Bekker, H., Modell, M., and Denniss, G. et al., 1993. 'Uptake of cystic fibrosis testing in primary care: supply push or demand pull?', *BMJ*, 306, pp. 1584-6.

Bradley, D.M., Parsons, E.P., and Clarke, A.J., 1993. 'Experience with screening newborns for Duchenne muscular dystrophy in Wales', *BMJ*, 306, pp. 357-60.

Bush, J.W., Chen, M.M., and Patrick, D.L., 1973. 'Health status index in cost effectiveness: analysis of PKU program', in R.L. Berg (ed). *Health status indexes. Chicago: Hospital Research and Education Trust*, pp. 172-209.

Chadwick, R., Levitt, M., and Shickle, D. (eds.), 1997. *The Right to Know and the Right Not to Know*, Avebury, Aldershot.

Chapple, J.C., Dale, R., and Evans, B.G., 1987. 'The New Genetics: Will it pay its way?', *Lancet*, pp. 1189-92.

Clarke, A., 1991 (a). 'Is non-directive genetic counselling possible?', *Lancet*, 338, pp. 998-1001.

Clarke, A., 1991 (b). 'Genetics, Ethics and Audit', *Lancet*, 335, pp. 1145-1147.

Clarke, A., 1993. 'Response to: 'What counts as success in genetic counselling?', *J Med Ethics*, 19, pp. 47-49.

Clothier Report, 1992. *Report of the Committee on the Ethics of Gene Therapy (Clothier Report)*, Cm.1788, HMSO, London.

Cochrane, A.L., and Holland, W.W., 1971. 'Validation of screening procedures', *Br Med Bull*, 27, 1, pp. 3-8.

Cuckle, H.S., and Wald, N.J., 1984. 'Principles of screening', in Wald NJ (ed.). *Antenatal and Neonatal Screening*, Oxford University Press, Oxford.

Department of Health Economics and Operational Research Division, 1994. *Register of Cost-Effectiveness Studies,* Department of Health Economics and Operational Research Division.

Dhondt, J-L., Farriaux, J-P,. Sailly, J-C., and Leburn, T., 1991. 'Economic Evaluation of Cost-Benefit Ratio of Neonatal Screening Procedure for Phenylketonuria and Hypothyroidism', *Journal of Inherited Metabolic Diseases*, 14, pp. 633-639.

Edgar, A., Salik, S., Shickle, D., and Cohen, D., 1998. *The Ethical QALY*, Euromed Communications , Haslemere.

Elborn, J.S., Shale, D.J., and Britton, J.R., 1991. 'Cystic fibrosis: current survival and population estimates to the year 2000', *Thorax*, 46, pp. 881-885.

Elsas, L.J., 1990. 'A Clinical Approach to Legal and Ethical Problems in Human Genetics', *Emory Law Journal*, 39, pp. 811-53.

Fulford, K.W.M., 1989. *Moral theory and medical practice*, Cambridge University Press, Cambridge.

Gill, M., Murday, V., and Slack, J., 1987. 'An economic appraisal of screening for Down's syndrome in pregnancy using maternal age and serum alphafetoprotien concentration', *Soc. Sci. Med.*, 24, pp. 725-31.

Hartley, N.E., Scotcher, D., and Harris, H., et al., 1997. 'The Uptake and acceptability to patients of cytic fibrosis carrier testing offered in pregnancy by the GP', *BMJ*, 34, pp. 459-64.

Holland, W.W., and Stewart, S., 1990. *Screening in Health Care: Benefit or bane?*, The Nuffield Provincial Hospitals Trust, London.

Holloway, S., and Brock, D.J.H., 1994. 'Cascade testing for the identification of carriers of cystic fibrosis', *Journal of Medical Screening*, 1, pp. 159-164.

Jones, S., and Shickle, D., et al., 1988. 'Acceptability of Antenatal Diagnosis for Sickle Cell Disease among Jamaican Mothers and Female Patients', *West Indian Medical Journal*, 37, pp. 12-15.

Kelves, D.J., 1985. *In the Name of Eugenics*, Harvard University Press, Cambridge (Mass.).

Kerem, E., Corey, M., and Kerem, B-S, et al, 1990. 'The relationship between genotype and phenotype in cytic fibrosis - Analysis of the most common mutation (ΔF508)', *New England Journal of Medicine*, 323, 22, pp. 1517-1522.

Kromrower, G.M., 1984. 'Phenylketonuria and other inherited metabolic defects', in Wald NJ (ed.). *Antenatal and Neonatal Screening*, Oxford University Press, Oxford.

Lappé, M., Gustafson, J., and Roblin, R., 1972. 'Ethical and social issues in screening for genetic disease', *The New England Journal of Medicine*, 286, 21, pp. 1129-1132.

Last, J.M., 1988. *A Dictionary of Epidemiology* (2 ed.), Oxford University Press, New York.

Lerman, C., Gold, K., and Audrain, J. et al., 1997. 'Incorporating biomarkers of exposure and genetic susceptibility into smoking cessation treatment: effects on smoking-related cognitions, emotions, and behavior change', *Health Psychology*, 16, 1, pp. 87-99.

Livingstone, J., Axton, R.A., and Gilfillan, A. et al., 1994. 'Antenatal screening for cystic fibrosis: a trial of the couple model', *British Medical Journal*, 308, pp. 1459-62.

Mennie, M.E., Gilfillan, A., and Compton, M. et al., 1992. 'Prenatal screening for cystic fibrosis', *Lancet*, 340, pp. 214-6.

Ministry of Health and Social Affairs (Norway), 1993. *Biotechnology related to Human Beings,* Ministry of Health and Social Affairs.

Modell, B., and Kuliev, A.M., 1991. 'Services for Thalassaemia as a Model for Cost-Benefit Analysis of Genetics Services', *Journal of Inherited Metabolic Disease*, 1, 14, pp. 640-651.

Mooney, G., 1992. *Economics, Medicine and Health Care* (2nd ed.), Harvester Wheatsheaf, New York.

Murray, J., Cuckle, H., Taylor, G., and Hewison, J., 1997. 'Screening for fragile X syndrome', *Health Technology Assessment*, 1, 4, pp. 1-71.

National Institutes of Health Consensus Development Program, 1997. *Breast Cancer Screening for Women Ages 40-49,* 103, pp. 21-23.

Nuffield Council on Bioethics, 1993. *Genetic Screening Ethical Issues*, Nuffield Council on Bioethics, London.

O'Brien, B., 1986. *'What are my chances doctor?' - A review of clinical risks*, Office of Health Economics, London.

Payne, Y., Williams, M., and Cheadle, J. et al., 1997. 'Carrier screening for cystic fibrosis in primary care: evaluation of a project in South Wales', The South Wales Cystic Fibrosis Carrier Screening Research Team, *Clinical Genetics*, 51, 3, pp. 153-63.

Phin, N., 1990. *Can economics be applied to prenatal screening?*, Centre for Health Economics, University of York, Discussion Paper 74.

Post, S.G., Botkin, J.R., and Whitehouse, P., 1992. 'Selective Abortion for Familial Alzheimer Disease?', *Obstetrics and Gynaecology*, 79, pp. 794-8.

Rommens, J.M., Iannuzzi, M.C., and Kerem, B. et al., 1989. 'Identification of the Cystic Fibrosis Gene: Chromosome Walking and Jumping', *Science*, pp. 1059-1065.

Rosenberg, S.A., Aebersold, P., Cornetta, K., Kasid, A., Morgan, A.A., and Moen, R., et al., 1990. 'Gene transfer into humans: immunotherapy of patients with advanced melanomna using tumor infiltrating lymphocytes modified by retroviral gene transduction', *N Engl J Med*, 323, pp. 570-8.

Russell, S.J., 1997. 'Gene therapy', *BMJ*, 317, pp. 1289-92.

Shickle, D., and Chadwick, R., 1994. 'The Ethics of Screening: Is 'screeningitis' an incurable disease?', *Journal of Medical Ethics*, 20, pp. 12-18.

Shickle, D., and Harvey, I., 1993. 'Inside-out, back-to-front: a model for clinical population genetic screening', *Journal of Medical Genetics*, 30, pp. 580-582.

Shickle, D., and Harvey, I., 1996. 'Screening for Cystic Fibrosis Carrier Status', in H.I.J. Wildschut, C.P., Weiner, T.J., Peters (eds.), *When to screen in Obstetrics and Gynecology*, WB Saunders Company Limited, London, pp. 87-99.

Shickle, D., and Harvey, I., 1996. 'Screening for Cystic Fibrosis Carrier Status', in Wildschut HIJ, Weiner CP, Peters TJ (eds.), 1996. *When to screen in Obstetrics and Gynecology*, WB Saunders Company Limited, London.

Standing Medical Advisory Committee, 1990. '*Blood Cholesterol Testing: The cost effectiveness of opportunistic cholesterol testing*', HMSO, London.

Strong, C., 1997. 'Ethics in Reproduction and Perinatal Medicine: A new framework', Yale University Press, New Haven, pp. 135-157.

Super, M., Schwarz, M.J., Malone, G., Roberts, T., Haworth, A., and Dermody, G., 1994. 'Active cascade testing for carriers of cystic fibrosis gene', *BMJ*, 308, pp. 1462-7.

The Cystic Fibrosis Genetic Analysis Consortium, 1994. 'Population variation of common cystic fibrosis mutations', *Human Mutation*, 4, pp. 167-77.

Tyler, A., Ball, D., and Craufurd, D., 1992. 'Presymptomatic testing for Huntington's disease in the United Kingdom', *BMJ*, 304, pp. 1593-6.

Wald, N.J., George, L.M., and Wald, N.M., 1993. 'Couple screening for cystic fibrosis', *Lancet*, 342, pp. 1307-8.

Wald, N.J., Kennard, A., Hacksaw, A., and McGuire, A., 1998. 'Antenatal screening for Down's syndrome', *Health Technology Assessment*, Vol 2, 1.

Walters, S., Britton, J., and Hodson, M.E., 1993. 'Demographic and social characteristics of adults with cystic fibrosis in the United Kingdom', *British Medical Journal*, 306, pp. 549-52.

Watson, E.K., Mayall, E., and Chapple, J., et al., 1991. 'Screening for carriers of cystic fibrosis through primary health care services', *BMJ*, 303, pp. 504-7.

Wertz, D.C., and Fletcher, J.C., 1988. 'Attitudes of Genetic Counsellors: A Multinational Survey', *American Journal of Genetics*, 42, pp. 592-600.

Williamson, R., Allison, M.E., and Bentley, T.J., et al., 1989. 'Community attitudes to cystic fibrosis carrier testing in England: a pilot study', *Prenat-Diagn*, Oct; 9(10), pp. 727-34.

Wilson, J., and Jungner, G., 1968. 'Principles and practice of screening for disease', *Public Health Papers*, WHO No 34, Geneva, WHO.

Royal College of Physicians, Working group of the Clinical Genetics Committe of the Royal College of Physicians, 1991. *Purchasers' Guidelines to the Genetic Services in the NHS*, Royal College of Physicians, London.

Table 1.1. Birth frequency of some genetic disorders in the UK (Royal College of Physicians, 1991)

Name of condition	Birth frequency
Autosomal dominant disorders	
Huntington's chorea	1 in 3,000
Familial polyposis coli	1 in 8,000
Adult polycystic disease of kidneys	1 in 1,000
Familial hypercholesterolemia	1 in 500
Tuberous sclerosis	1 in 12,000
Neurofibromatosis	1 in 2,500
Von-Hippel Lindau disease	1 in 100,000
Retinitis pigmentosa	1 in 5,000
Bilateral retinoblastoma	1 in 30,000
Myotonic dystrophy	1 in 7,000
Autosomal recessive disorders	
Cystic fibrosis	1 in 2,000
Adrenal hyperplasia	1 in 10,000
Friedreich's ataxia	1 in 54,000
Spinal muscular atrophy	1 in 10,000
Phenylketonuria	1 in 13,000
Usher's syndrome	1 in 27,000
Sickle cell disease in Afro-Caribbeans	1 in 250
Thalassaemia	
in Cypriots	1 in 140
in Indians	1 in 1000
in Pakistanis	1 in 300
X-Linked recessive diseases	
Duchenne/Becker muscular dystrophy	1 in 9,000
Haemophilia	1 in 20,000
X-Linked retinitis pigmentosa	1 in 7,000
Other X-linked eye disorders	1 in 7,000
Fragile X syndrome	1 in 4,000
Other forms of X linked severe mental disorders	1 in 4,000
Chromosomal disorders	
Unbalanced translocations	1 in 2,000

Table 1.2. Comparison of screening strategies at different times in life cycle (Shickle and Harvey, 1996, p. 90)

	Advantages	Disadvantages
Neo-natal screening	Mechanisms already exist for collection of blood samples Community midwife/health visitors provide counselling Early treatment possible	Autonomy increased in subsequent pregnancies only Parents give consent on behalf of the child Knowledge that child not "perfect" can affect parent/baby relationship Knowledge of carrier status only of importance for the next generation Information may be lost/misunderstood before child reaches reproductive age Problem of detection of non-paternity or sex chromosome abnormalities
Ante-natal screening	Easy to organise Most pregnant women attend ante-natal clinic Performed along with other ante-natal tests Only offered to people of reproductive age More receptive to issues affecting their children	Consent may not be fully informed Availability of partner may be a problem Limited period to perform counselling etc. Anxiety about pregnancy may make counselling more difficult Termination may not be an acceptable option
Childhood (teenager) screening	Aware of being a carrier before long term relationship formed Could choose to avoid having children with another known carrier	Consent a problem Adolescence already a time of sexual confusion Stigmatisation
Young adults of reproductive age	Maximises autonomy and reproductive choice Informed choice easiest to obtain	Primary care staff may not have skills or time to devote to counselling Many young adults do not visit their GP Many pregnancies not planned May be a middle class bias if individuals have to request screening

Table 1.3. Possible outcomes of screening test

Disease

Finding		Present	Absent	
	Positive	True Positive A	False Positive B	A+B
	Negative	False Negative C	True Negative D	C+D
		A+C	B+D	

Sensitivity =   Proportion of patients with the disease in whom the finding is positive i.e. correctly identified by the test

$$= \frac{A}{A+C}$$

Specificity =   Proportion of those without the disease in whom the finding is negative i.e. correctly excluded by the test

$$= \frac{D}{B+D}$$

Table 1.4. Costs and benefits associated with being a true positive, false positive, false negative or true negative

True Positive		False Positive	
Costs	Benefits	Costs	Benefits
May be no increase in life expectancy if treatment is of no effect but patient may have to live longer with diagnosis because it is made earlier (lead time bias)	Earlier treatment (cheaper; more pleasant, better prognosis) Benefits of "sick role" (excused social responsibilities provided seek and comply with treatment)	Unnecessary anxiety and stigma until correct diagnosis is made Unnecessary investigations and treatment (time consuming; expensive; pain; side effects; complications)	Provides an opportunity to counsel on unhealthy lifestyles: "next time could be for real"
Anxiety: "Worried Ill" Stigma Reduced quality of life e.g. due to side effects or complications of treatment	Explanation of symptoms: knowing the cause of symptoms so that they can be "treated" can relieve anxiety	Lingering doubts: patients tend to believe doctors! may be difficult to reassure patient that initial diagnosis was incorrect.	

Table 1.4. (cont.)

False Negative		True Negative	
Costs	Benefits	Costs	Benefits
False reassurance ("Unworried Ill")	Spared anxiety if treatment of no benefit	Anxiety while awaiting results of test ("Worried Well")	Reassurance
May legitimise "Unhealthy lifestyle"		May legitimise "unhealthy lifestyle e.g. "if my chest x-ray is normal, then smoking cannot be doing me any harm!"	
Later treatment (unpleasant; more expensive; worse prognosis)			
Increased cost per desired outcome			

Table 1.5. Cuckle and Wald's (1994) requirements for a worthwhile screening programme

Aspect	Requirement
1. Disorder	Well-defined
2. Prevalence	Known
3. Natural history	Medically important disorder for which there is an effective remedy available
4. Financial	Cost-effective
5. Facilities	Available or easily installed
6. Ethical	Procedures following a positive result are generally agreed and acceptable both to the screening authorities and to the patient.
7. Test	simple and safe
8. Test performance	Distribution of test values in affected and unaffected individuals known, extent of overlap sufficiently small, and suitable cut-off level defined.

Chapter 2

Genetic screening, information and counselling in Austria

GERTRUD HAUSER
Institute of Histology and Embryology
University of Vienna
Austria

Since the majority of women who are pregnant deliver their babies in hospital, screening of newborn there presents no difficulties. Screening however is also carried out in those born at home, so that all babies are examined and tested. This, like other pre-natal, neo-natal and paediatric services is greatly facilitated by the mother-child passport into which the results of all examinations, tests, and inoculations are entered. Financial inducements, together with the normal maternal desire to act in the best interests of the child, ensure virtual 100% coverage - a mother receives a monetary grant when the child is born and at regular intervals afterwards provided the passport shows that she has attended for the required ante-natal and post-natal examinations and care checks up to the third year.

The effectiveness of these measures has contributed to the change in neo-natal mortality: before the introduction of the mother-child pass in 1972, there were 13 neo-natal deaths per thousand, today this figure has dropped to 5 per thousand.

1. ANTE-NATAL

The routine ante-natal tests (haematological: Wassermann, blood group incompatibility, rubella, toxoplasmosis, cytomegalovirus and ultrasound

35

R. Chadwick et al. (eds.), The Ethics of Genetic Screening, 35–41.

examination) are carried out on every mother. In addition, all mothers older than 35 years are offered free investigation for fetal chromosomal abnormalities, accompanied by genetic counselling. These investigations are also available for others at the request of the family doctor (general practitioner) or specialist, often prompted by the woman herself, on account of a previous medical history or family history.

The same route (via referral) is also open in the case of other genetic disorders. Although testing facilities are not available in all hospital centres, women in need from anywhere in the country are sent to those units which specialise in a particular disorder. For example individuals with a family history of Duchenne muscular dystrophy may be referred to the Institute for Human Genetics in Graz or in Vienna; or the Department of Human Genetics of the Wilheminen Hospital in Vienna, for specialist advice on the mucopoly saccharidases.

2.   NEO-NATAL

After birth, blood specimens are obtained from the babies for screening. The disorders sought are: phenylketonuria, its variants hyperphenylalaninemia and biopterin disturbances, and congenital disorders of thyroid output. Other conditions for which testing is available, though not on all births, are:. argininosuccinase deficiency, galactosemia, galactokinase deficiency, epimerase deficiency, phosphogluconnutase deficiency, biotin-idase deficiency, maple syrup urine disease, homocysinuria. All infants are screened by ultrasonography for congenital dislocation of the hip.

3.   SCREENING IN CHILDHOOD

The mother-child pass is the regular health investigation of the growing infant. This includes screening for disorders of hearing and vision, mental and physical development. There is no question of requiring parental consent; the parents may opt out from these investigations, but parental concern and money are strong incentives.

4.   SELF-HELP GROUPS

Self-help groups of varying sizes, largely consisting of patients and members of their families are mainly based in Vienna but with members scattered throughout the country. These groups are recognised by the hospitals which make facilities available for them to meet. Doctors and other members of hospital staff assist with lectures and discussions. There are groups devoted to: retinitis pigmentosa, Duchenne muscular dystrophy, Down's syndrome, phenylketonuria (PKU), deafness, mucopolysaccharidosis, Cystic fibrosis.

5.   GENETIC COUNSELLING

Official genetic counselling centres supported by the national health services which meet the salaries of doctors, nurses, laboratory technicians, clerical staff and running expenses, exist in a number of hospital centres. Altogether there are nine of these centres, although others are planned.

Vienna:	Wilhelminenhospital: Department of medical genetics. University hospital: Department of medical genetics (Internal medicine), Department of pre-natal diagnosis and therapy (obstetric clinic), Department for pre-natal diagnosis (paediatric clinic), Department of social psychiatry (psychiatric clinic)
Graz:	Institute for Human Genetics
Uinz:	Institute for Genetic counselling (provincial paediatric clinic)
Salzburg:	Institute for Genetic counselling (provincial paediatric clinic)
Innsbruck:	Institute for Medical Biology and Human Genetics

6.   SENSITIVE ISSUES

The majority of the ethical concerns raised in other European countries (e.g. requirements of insurance companies for genetic testing, informed consent, screening of children) have not received the widespread discussion in Austria that they have elsewhere. Of particular relevance is the question of genetic registers. There is still a strong memory of the misuses to which personal details were put during the national socialism years. So that there is a strong antipathy to any form of genetic register being kept. For the sake of the patients, clinics and individual doctors naturally keep confidential case records, but there is no formal policy of maintaining registers at genetic centres.

7.   LEGAL POSITION

The legal position of genetic screening, investigations and related matters are defined in Federal laws. Federal law 510 (gene-technique Law - GTG) relates to the application of genetic techniques and covers aspects relevant to controlling human reproduction. It was issued in the Federal law publication for the Austrian Republic on July 12th 1994 (pages 4111-41 50).

It regulates work with organisms changed by genetic techniques (GVO), the freeing and distribution of GVO and the application of gene analysis and gene therapy to man, and also modifies a previous decree dealing with responsibility for products.

The general aims of the law are :

1. to protect health of man and his descendants from damage which may be caused a. directly by manipulation of the human genome, by gene analysis of man, or by effects on man from GVO, b. indirectly by effects on the environment from GVO, and also to protect the environment (especially ecosystems) from noxious effects of GVO and thus guarantee a high amount of security for man and the environment.
2. to support application of gene techniques for the sake of man's well-being by creating a legal frame for their research, development and use.

This federal law applies for:

1. gene technique institutions;
2. work with GVO,
3. freeing of GVO;
4. distribution of products consisting of or containing GVO;
5. labelling of products consisting of GVO or of parts of them or have been produced from them It does not lead to GVO as in gene analysis and gene therapy in man, especially with:
    1. In vitro fertilisation,

2. Conjugation, transduction, transformation or any other natural process,
3. Polyploidy induction and elimination of chromosomes,
4. procedures of undirected mutagenesis,
5. Cell and protoplast fusion,
6. production of somatic - human or somatic animal Hybridoma cells
7. Self-cloning of non-pathogenic naturally existing microorganisms.

Section IV of the law is of special relevance to Gene analysis and gene therapy in man. It prohibits intervention in human reproductive cell lines and specifies the conditions under which gene analysis may be done. Gene analysis in man for medical purposes may only be done at the request of a physician specialised in Human Genetics or by a physician specialised in the relevant clinical field for:

a. verification of a predisposition for a disease, especially for predisposition of a possibly later onset hereditary disease, or
b. verification of a carrier, or at the request of the treating physician or the physician responsible for making the diagnosis for a. diagnosis of a manifest disease or a connected disease possibly manifesting later, or b. preparation of a therapy and monitoring its effect, or c carrying out necessary investigation of a relative(s).

Written consent for a gene analysis must be given by the patient setting out that she (he) had been previously informed by a physician or specialist physician about the type, extent and informativeness of gene analysis. In the course of a pre-natal investigation however, a gene analysis may only be done as far as medically necessary with the written consent of the pregnant woman certifying that she had been previously informed by a physician or specialist physician about the type, extent and informativeness of gene analysis, and the risks of the proposed intervention. For a minor or for a person not considered officially responsible for herself this certification has to be given by a legal guardian or by the legal representative designated for such, whose jurisdiction also includes gene analysis.

Before and after carrying out gene analyses to detect disposition for a hereditary disease or carrier status the requesting physician must carry out extensive counselling of the person to be investigated; if this gene analysis is done in the course of a pre-natal investigation in a pregnant woman incompetent to give her consent, her legal guardian or designated representative responsible for the consent must also be present.

The counselling has to include the complete relevant discussion of all results of investigation and medical facts as well as their social and psychological consequences, and must not occur in a directive way in the case of pre-natal gene analysis. Usefulness of additional counselling by a non medical psychotherapist or social worker has to be pointed out information conveyed in the counselling is to be made available in a written form. The initiator for gene analysis has to recommend to the person investigated to advise possibly affected relatives to seek genetic counselling, if the results of the gene analysis require the inclusion of a relative for verification or there appears to be serious risk of the disease in relatives of the investigated persons.

Whoever performs analysis or orders it has to keep the data confidential. Full information on all the data must be given to the person investigated if requested. Unexpected results of direct clinical relevance or which the investigated person had expressly asked for have to be given to her, but in such a way as not to cause distress. In borderline cases this information need not be disclosed.

Data that are not anonymous may only be used for any purpose other than the original cause of the investigation only with express and written consent of the person investigated.

Data may only be given to: a. persons who are directly concerned with collection, analysis or computation of data in the institution where they were collected, b. the person investigated, c. persons giving consent on behalf of the patient, d. the physician who requested the gene analysis, to the treating physician or the physician making the diagnosis, e. to others only insofar as the investigated person has given express and written consent, and has the possibility of a written withdrawal.

Data have to be protected from unauthorised access.

Data that are not anonymised may only be computerised in the institution in which they were taken and by the physician who requested gene analysis; they have to be stored separately from other data and may only be called up by persons who have access according to this federal law and only with a special permit.

Gene analyses on man for medical purposes may only be done in institutions licensed for it. A licence requires application by the director of the institution to the federal Ministers of Health, Sports, and Consumer Protection. In deciding whether to issue a licence, the minister takes into account advice from the appropriate scientific council regarding the facilities of the institutions and the state of knowledge and technique. He has to grant a license if the appropriate scientific council states that the facilities and personnel of the institution meet the required level of science and technique and data protection. He has to withdraw the licence if these requirements cease to be met, or if gross deficiencies in the service occur, withdraw it temporarily till these have been rectified.

Gene analyses in man for scientific purposes and for training may only be done on the express and written consent of the donor of the sample or on anonymised samples.   A sample used for scientific purposes is also considered anonymous if it is only identified by a code in the relevant institution and not with the name of the donor. Results from gene analysis may only be distributed or published if correct measures have been taken to ensure that the donor cannot be identified (except through the relevant institution).

Employers and insurance companies, also including their representatives and collaborators, are forbidden to obtain, request, take or otherwise use results of gene analyses of their employees, applicants for employment, or for those who wish to insure themselves.

8.   CONCLUSION

From the foregoing it appears that Austria is keeping pace with other European countries in the development of genetic services. As elsewhere this development relies on the interest of individual professionals for the establishment of specific testing. As regards the legislation it is possible that Austria is ahead of others.

Chapter 3

The Belgian perspective on genetic screening

KRIS DIERICKX
Centre for Biomedical Ethics and Law
Catholic University of Leuven
Belgium

1. INTRODUCTION: THE BELGIAN CONTEXT

Belgium has eight centres for human genetics which are involved in fundamental research on different aspects of the human genome. All eight centres for medical genetics belong to a university and are funded by the Ministry of Public Health. The research is supported within the normal university budget of the centres. These limited financial resources available to geneticists in a small country like Belgium necessitate, both on the scientific level as well as on the political level, an obligation to look with great interest to the policy of the Commission of European Communities regarding human genome analysis. Most of the centres for human genetics are offering clinical services. In some of them there is a well elaborated counselling team which offer both genetic counselling and fundamental research.

Until now most of the patients find their way to those university centres for human genetics by referral from their physician, by one of the patient associations or by referral from smaller hospitals. Everyone may freely ask for genetic counselling and patients only have to pay a small part (£7) of the real costs (£200) for the genetic tests. In principle there are no special conditions or any barriers for consultation and diagnosis. No matter how complicated and long the procedure might be, those persons involved incur no extra costs. The centres have a good counselling reputation which might

R. Chadwick et al. (eds.), The Ethics of Genetic Screening, 43–53.

be one of the reasons why there is little public debate on the topic of genetic screening. Patients, medical doctors and patient associations are content with the services offered by the genetic (counselling) teams. This is a positive point: in Belgium there is a high degree of confidence in the medical profession. This could explain in part the lack of discussions and laws on this topic. On the other hand, this remains a very fragile situation.

There are no government statements or official policy documents on the ethical or other aspects of genetics, genetic screening or on bioethics in general. In practice however (non-genetic) screening is being done in Belgium. For example, private companies are screening their employees for illicit drug abuse, mostly without consent. On the other hand, it is worthy to note that some (non-genetic) screening like that of AIDS requires prior consent as stated explicitly by the National Council for the Order of Physicians. But AIDS is not a genetic disorder. One province has recently started a screening program for cystic fibrosis and some muscular diseases.

As a conclusion to this introduction, we could say that until recently the public debate in the Belgian society on genetic screening was non-existent or still in its infancy and mainly confined to small, mostly scientific circles.

2.   RECENT EVOLUTIONS

The mid-nineties demonstrated a turning point in this rather 'quiet' domain. An acceleration in the public's encounter with information on genetics is evident from several sides: government, social organisations, scientific community.

In January 1996, a Federal Council for Bioethics was established (Nys, 1996). Although there are at this moment no workgroups on genetics or genetic screening, this may become possible in the future.

Until the first half of the nineties, the only governmental initiative in this whole field was a national scientific symposium in 1987 on the ethical and social aspects of contemporary developments in medicine. One of the preparatory working groups dealt explicitly with human genetics, but the only result was a vague formulation of some very general ethical questions and remarks in the margin of a presentation of biological, medical, genetic and clinical information (De Meester, 1987).

In March 1997 a scientific conference was held on the topic of prevention and health promotion (XXX, 1997). One of the seven workgroups examined the subject of predictive genetic medicine. Some fifteen scientists were involved in this workgroup. There was also one discussion group examining screening (genetic and non-genetic). This Conference aimed at reaching a wide range of the population through a highly visible presence of the meeting itself and by the resulting media coverage.

The Flemish government also sponsored a widespread brochure on human genetics: 'Focus on genetics' (Fryns, 1996).

Another new evolution is that social cultural organisations began to invite speakers on the topic of genetics, gene therapy and genetic screening. One large organisation began in 1995 registering more than 200.000 female members. They collaborated with scientists producing a three year long information and education campaign for its members: the first year the medical genetic aspects were discussed, the second year the psychosocial, and in the last year the ethical aspects.

In 1996 the K.U.Leuven organised an elaborate exhibition on human genetics: 'The ABC of the DNA: man and genetics' (Marynen, 1996). People visited this one month exhibition from all over the country and several television programmes have focused on it. The Centre for Human Genetics in Leuven has eleven different videotapes available: one with basic information about genetics, genetic risks, genetic tests, destined for schools and education. Other videotapes relate to specific genetic diseases, such as Prader-Willi syndrome, Fragile-X, Turner syndrome, Down's syndrome, Angelman syndrome, etc. Almost all of the tapes are available in French, English and Dutch. This collective effort hopefully will result in a better informed Belgian population

3. ACTUAL KNOWLEDGE OF THE BELGIAN POPULATION ABOUT HUMAN GENETICS

3.1 General population

The permanent progress in medical genetics creates a continual accumulation of a mass of information. Only a limited group of genetic specialists however is sufficiently acquainted with the present and future clinical applications in this field to begin to understand their implications. A growing gap emerges between the knowledge of this eclectic group of experts and the knowledge of the general public, who are the potential users of these new diagnostic tools. Carrier screening, predictive testing, amniocentesis, and other medical procedures are available to diagnose genetic disorders such as Down's syndrome, cystic fibrosis, etc. The decision making process regarding the use of these diagnostic possibilities should be based on sufficient information. Minimal knowledge about the tests, about the risks of developing a genetic disease and/or having an affected infant and about the nature and seriousness of common genetic diseases and congenital malformations is a prerequisite for a free and informed choice about a medical procedure. What do people know about genetics? Are they aware of the availability of clinical genetic services and genetic screening?

Research projects (Decruyenaere, 1992a) inquiring about knowledge of the frequency of genetic diseases and congenital malformations in newborns show that 31% of the respondents (n=385) know that the general risk for having an affected child is 3 to 5%. About one third underestimate this risk and another third overestimate the risk. Down's syndrome was the best known: almost everybody (97%) had heard about it. It is as well known as AIDS. A relatively large proportion are also familiar with the terms haemophilia and Spina Bifida: 76% and 79% have heard of it, respectively. Less well known was cystic fibrosis (58%) and the fragile X syndrome was unknown: 93% had never heard of it. These results indicate that the group they studied had some awareness of genetics, however with serious gaps: people are insufficiently acquainted with genetics. They observed a strong association of the knowledge of basic genetic information with the educational background of the respondents and with their professional activities. Persons with a higher educational background or persons who work(ed) with ill or handicapped people in a professional context have more knowledge as compared to individuals with a lower educational level or without these contacts. Another factor affecting knowledge of genetics was the age of the respondents: younger persons are more acquainted with basic genetic information and with methods of pre-natal diagnosis than older ones are. An increase of basic knowledge in the general Belgian population is essential for informed decisions about genetic screening.

3.2   Relatives of patients: the example of cystic fibrosis

The knowledge of parents of a patient with a genetic disorder about that specific disease and its genetic transmission tends to be much better than the knowledge of the general population. A Belgian study showed that 87% of the parents of a patient with cystic fibrosis (CF) knew their 1 in 4 recurrence risk at each subsequent pregnancy (Denayer, 1990).   The probability at that specific relatives of the CF-child are carriers of the CF gene was less well known. The probability that brothers and sisters of the parents would be carriers of the CF-gene was known by only 18% of the respondents. The parental knowledge regarding the autosomal recessive transmission mode of CF was primarily acquired by getting information from the paediatricians treating their child. A substantial contribution was also made by the information disseminated by the Belgian Cystic Fibrosis Association. Although most aunts and uncles of a CF-patient are rather well informed about the main symptoms of CF, they have only a superficial knowledge about the genetic transmission of this disease (Denayer, 1992). This may be partially explained by the fact that there is still a taboo on hereditary diseases; partially by the gaps in the knowledge of the parents and the relatives.

4. ATTITUDES TOWARDS GENETIC SCREENING

4.1 Neo-natal screening

Neo-natal screening in Belgium has been limited thus far to one treatable disorders, namely phenylketonuria, some metabolic diseases and cystic fibrosis (in one province). Blood is obtained from the babies (three days old) by heel prick, mostly without (informed) consent or sometimes even without the communication that it will be done. The parents are supposed to know that the test is taking place.

4.2 Carrier screening: the case of cystic fibrosis

Carrier screening can identify individuals or couples with a high risk of having a child suffering from a specific recessively inherited disorder. It could prevent the birth of the first (and following) handicapped child in all families at risk. Indeed, couples at risk could decide not to have children, or to make use of pre-implantation or pre-natal diagnosis for the specific disorder.

In Belgium there was a research project to evaluate the attitudes of the general population with regard to the attitudes toward carrier screening for cystic fibrosis (Decruyenaere, 1992b). After the participants (n=385) had read the information, they were asked whether they wanted to know their carrier status for the CF-gene. Sixty-three percent expressed interest in knowing it. The results are independent of age, number of children one has, planning of pregnancies at the time of the survey, educational level, religious convictions, perceived burden of CF, having heard of CF, knowledge score for CF and perceived susceptibility for genetic diseases. They also asked respondents' opinion of community-wide carrier screening program for CF. Nine percent agreed with systematic screening, organised by the government. Receiving sufficient information so that they are free to decide whether they take the test or not was preferred by most individuals (76%). Nine percent reported that they only want to receive information on screening when asking for it. No one held the opinion that screening is unacceptable and only 1% perceived it as not necessary.

An identical question with the same alternatives was posed with regard to carrier screening for recessive disorders in general. The results gave the same picture as for screening for CF. The attitude towards carrier screening was not associated with age, religious convictions, educational background, knowledge of CF, knowledge of basic genetic information nor with perceived susceptibility for genetic diseases. There was a rather good consensus about the optimal moment for carrier screening. The majority held the opinion that screening programme should be available prior to pregnancy: 46% advocated testing before marriage and 40% before pregnancy. Only 2% would choose testing at the beginning of the pregnancy. One of the major conclusions is that the group selected from the Belgian population seem to accept voluntary pre-pregnancy carrier screening. Although the majority had no objections against carrier screening, at least "when it would not be imposed by the government", the proportion of respondents expressing interest in knowing their own carrier status is slightly lower.

4.3   Predictive testing: the example of Huntington's disease

This type of 'screening' would be directed at identifying persons with genetic pre-dispositions for specific common late(r) onset disorders. Predictive testing for Huntington's disease is considered to be a test case for other late onset diseases.

4.4 Predictive testing for at-risk persons (before the availability of the test)

In 1984 a Belgium study investigated the opinion of 49 at-risk persons and 27 partners of at-risk persons (Evers-Kiebooms, 1987). About the same percentage of persons at risk (57%) and partners (60%) would want to take the predictive test. The main difference between those at risk and their partners was that the partners were in favour of taking the test immediately after it was available, while the at-risk persons preferred to postpone the test. For the group of persons at risk, there was an association with the age at onset of Huntington's disease in the affected parent: the younger the parent of a subject at risk became affected, the stronger the desire to take the test seemed to be. For the partners the presence or absence of children played an important role in their attitude towards predictive testing. This association was not found in the group with the at-risk persons.

A larger survey by the same research group in 1987 also investigated the attitudes of at-risk persons and partners toward predictive testing (Evers-Kiebooms, 1989). As compared to the first survey, more persons were in favour of taking the test. Sixty-six percent of the at-risk subjects and 74% of their partners wanted to have the test performed. Relatively more partners than at-risk persons were in favour of taking the test immediately when available, but the difference was not statistically significant. This study also showed a greater inclination to take the test if age of onset in the parent was younger. When we look at the reasons for deciding in favour or against taking the test, the data draws our attention to important differences in the motivation of at-risk persons on one hand and the partners on the other hand. At-risk persons tend to be especially concerned about their own future, while partners in addition tend to be concerned to the same extent about children (to be born or already born) and about the consequences for the relationship.

4.5 General population (test available)

In a more recent study, opinions on the implications of predictive testing for Huntington's disease were evaluated in a group of 169 women (between 21 and 35) with no special pre-existing knowledge or training in genetics (Decruyenaere, 1993). In the hypothetical situation of having a 50% risk for developing Huntington's disease, about half of the group (53%) expressed interest in a predictive test, while 25% answered that they were not interested and 16% did not know. Of those who accepted pre-natal diagnosis, 44% would terminate the pregnancy if the fetus had the late onset disease. This means that the other 56% implicitly agreed with the testing of minors but probably these persons did not consider the implications of the abortion option. The attitudes towards predictive testing were systematically associated with perceiving 'having more certainty about the future' as an advantage and with rejecting 'knowing everything in advance' as a disadvantage.

Since the detection of the gene in April 1993, there was a significant increase of at-risk persons who wanted to undergo the test in the Centre for Human Genetics in Leuven. In the period between November 1987 and April 1993, 125 persons asked for a test. Five of them, being minors, were refused. Persons who want to undergo the test since April 1993 test, are mostly persons other than those who asked for a test before the detection of the gene.

5.   SOCIAL ASPECTS OF GENETIC SCREENING

The general taboo on genetic disorders, leaves the research on the social implications of genetic screening fraught with difficulties. A study on the implications of being identified as a carrier of the CF-gene clearly shows that carriership of the CF-gene has negative connotations (Evers-Kiebooms, 1994). The selected group as a whole attributed more negative feelings to most carriers of the CF-gene than to most non-carriers. Non-carriers do think in a more negative way about 'most carriers' than do the carriers themselves. Moreover, the negative connotations of CF-carriership are stronger in those who have no personal experience with CF.

It is assumed that similar negative connotations about carriership will exist in the general population, and that the less one knows about the carrier condition, the stronger the prejudices will be. It is hypothesised that the information given by professionals will have a major impact in increasing or decreasing the stigmatising effect of carriership of CF. When we consider the respondent's self-description, it is clear that carriers of the CF-gene have significantly less positive feelings about themselves than non-carriers, and that women have more intense feelings, more positive when non-carrier and more negative when carrier. Most interesting, however, is the finding that carriers have significantly more positive feelings about themselves than they think 'most carriers' would have. This seems to suggest that the individual carrier, when hearing the 'bad news' about himself, is able to counterbalance the threat by a psychological defence mechanism, in order to safeguard his self-image (c.f. the phenomenon of 'illusory superiority').

6. LEGAL ASPECTS OF GENETIC SCREENING

There have been very few legislative developments regarding genetic screening. Normally, the general principles (e.g. informed consent, human rights) are applied or for specific situations and thus one tries to come to a form of self-regulation (cfr Huntington Association). Indirectly, we can find something about genetic testing and about genetic information (Nys, p. 10.15.37).

In Belgium a total ban on the use of genetic testing to predict the future health status of applicants for (life) insurance was laid down in the Law on Insurance Contracts, which came into force in September 1992. Article 95, subsection 1 on medical information and private insurance lays down that medical examination necessary to establish and fulfil the contract:

> may only depend on the anamnesis of the present health condition of the candidate and not on genetic research techniques which are meant to determine the future health condition.

A complete prohibition on communication of genetic data to insurers without any exception was laid down in the same Law on Insurance Contracts. Article 5, subsection 1 which applies to all insurance contracts, requires the insurance taker to give accurate information of all known circumstances of which he can reasonably assume to aware that may influence the risk-assessment performed by the insurer, though he is not obliged to give information on circumstances the insurer already knows or within reason should have known. He is not allowed to give genetic information spontaneously. The complete ban on giving genetic information includes information given by physicians, insurance takers, and insurers. This also means that the insurance taker is not allowed to volunteer favourable genetic information to get better conditions and lower premiums. This is to ensure the avoidance of any discrimination between insurance takers with genetic 'good luck' or 'bad luck'.

7.   CONCLUSION

The dominant moral principles concerning genetic counselling are considered to be the client's right to information, his/her right to confidentiality, and the principle of non-directive counselling. Given research results, it is difficult to say that these conditions have been fulfilled for genetic screening in Belgium. The general genetic knowledge was very poor, which is not a good starting point for ensuring that consent is 'informed'. Since the mid-nineties a joint effort from the government, the scientific world, and social organisations is attempting to improve the genetic knowledge of the whole population and to prepare the citizens for the genetic challenges of the near future.

REFERENCES

Decruyenaere, M. et al., 1992a. 'Community knowledge about human genetics, Birth Defects, *Original Article Series*, 28,1, pp. 167-184.

Decruyenaere, M., et al., 1992b. 'Cystic fibrosis. community knowledge and attitudes towards carrier screening and pre-natal diagnosis', *Clinical Genetics*, 41, pp. 189-196.

Decruyenaere, M. et al., 1993. 'Perception of predictive testing for Huntington's disease by young women. preferring uncertainty to certainty?', *Journal of Medical Genetics*, 30, pp. 557-561.

De Meester-De Meyer, W. (ed), 1987. *'Bio-ethica in de jaren '90 [Bioethics in the nineties]*, Omega, Gent.

Denayer, L. et al.., 1990. 'A child with cystic fibrosis. I. Parental knowledge about the genetic transmission of CF and about DNA-diagnostic procedures', *Clinical Genetics*, 37, pp. 198-206.

Denayer, L. et al.., 1992. 'The transfer of information about genetic transmission to brothers and sisters of parents with a CF-child, Birth Defects, *Original Article Series*, 28,1 pp. 149-158.

Evers-Kiebooms, G. et al.., 1987. 'Attitudes towards predictive testing in Huntington's disease. a recent survey in Belgium', *Journal of Medical Genetics*, 24, pp. 275-279.

Evers-Kiebooms, G. et al., 1989. 'The motivation of at-risk individuals and their partners in deciding for or against predictive testing for Huntington's disease', *Clinical Genetics*, 35, pp. 29-40.

Evers-Kiebooms, G. et al., 1994. 'A stigmatising effect of the carrier status for cystic fibrosis?', *Clinical Genetics*, 46, pp. 336-343.

Fryns, J.P. (ed.), 1996. *Erfelijkheid in de kijker [Focus on genetics]*, Ministerie van de Vlaamse Gemeenschap, Brussels.

Marynen, P. et al.., 1996, *Het ABC van het DNA. mens en erfelijkheid (The ABC of the DNA. Man and genetics)*. Davidsfonds, Leuven.

Nys, H. et al.., 1993. *'Predictive genetic information and life insurance. Legal aspects: towards European Community policy?'*, Rijksuniversiteit Limburg, Maastricht.

Nys, H., 1996. 'Preparation for the establishment of a Committee on Bioethics, *International Digest of Health Legislation,* 14, pp. 409-410.
XXX., 1997. *Preventive healthcare*, Kluwer Editorial, Diegem.

Chapter 4

Thalassemia prevention in Cyprus
Past, present and future

PANAYIOTIS IOANNOU
Cyprus Institute of Neurology and Genetics
Nicosia
Cyprus

As an increasing number of genetic disorders become preventable through the rapid increase in knowledge achieved in the course of the Human Genome Project, it may be useful to look at the example of prevention of thalassemia in Cyprus. Over the past 20 years, Cyprus has been able to develop one of the most effective prevention programmes for thalassemia or any other genetic disease, despite its relatively low level of scientific infrastructure and political misfortunes. The experiences gained in Cyprus may thus be of some relevance to the increasing number of preventable genetic disorders in general. These experiences may be especially relevant to many other developing countries, which are also gradually finding out that genetic disorders like thalassemia and sickle cell disease constitute a major health burden once the most serious aspects of starvation, infectious diseases, or war ravages as in the case of Cyprus, are contained.

This chapter provides an analysis of the main factors which have led to the success of the Cyprus thalassemia prevention programme and examines its impact on efforts to develop a coherent overall approach to the effective therapy of this devastating disease.

R. Chadwick et al. (eds.), The Ethics of Genetic Screening, 55–67.
© 1999 *Kluwer Academic Publishers. Printed in the Netherlands.*

1.   THE SITUATION IN CYPRUS PRIOR TO SCREENING

Although there is archaeological evidence from skeletal remains for the presence of thalassemia from ancient times (Angel, 1996), thalassemia was first recognised in Cyprus as a disease entity in 1947 by Dr. A. Faudry, during a post-war campaign to eradicate malaria from Cyprus. At that time most patients were dying in the first few years of life without any specialised support, although some patients with mild thalassemia intermedia survived to adulthood. Families of patients became socially stigmatised, a situation which has been largely overcome nowadays, although the word 'stigma' is still generally used in spoken language to denote the heterozygous state of thalassemia.

The only options available to parents in the early 1950s and 1960s were to watch their children suffer to death far from hospitals and the public eye, while praying for 'forgiveness and mercy'. The genetic nature of the disease was not well appreciated at that time by the general population and thus many couples had to go through the terrible ordeal of having to bury one or more of their children at the age of 2-10 years, in their efforts to have a healthy family. The high frequency of thalassemia in Cyprus meant that almost every extended family would have one or more patients in its branches.

Similarly, every small village would have a number of families stigmatised by thalassemia. There was thus a high awareness of the effects of the disease among the population, even though the presence of thalassemia patients would be a closely kept secret within the family in the bigger villages and towns.

Progress in knowledge about thalassemia in other countries filtered slowly through to Cyprus in the early 1960s and families started to seek a better treatment for their children. Despite the heavy stigma associated with the disease, many parents were courageous enough to go out searching for blood donors for their children, but most could hardly afford to pay for the precious blood through their meagre wages. This search for paid blood and the increased hospitalisation of the patients for blood transfusions, created the opportunities for unrelated families going through the same suffering to meet and to start forming the first parents pressure group. The formation of the Pancyprian Anti-anaemia Association in the late 1960s marked indeed a very significant step forward in the struggle against thalassemia, since it helped to focus attention on the plight of the patients and their families and on the social magnitude of the problem.

However, initial efforts to develop a government policy on thalassemia in the late 1960s and early 1970s fell victim to the political turmoil in Cyprus between the Greek-Cypriot and Turkish-Cypriot communities and to the Turkish invasion of Cyprus in 1974. Despite the immense difficulties of managing the consequences of the Turkish invasion and re-housing about 40% of the population, a prevention programme was put in place shortly after the Turkish invasion. This led to the establishment of the Cyprus Thalassemia Centre in 1978 and the inauguration of its own buildings in 1980, largely build with funding from the Holy Church of Cyprus.

2. IMPROVED MANAGEMENT OF THALASSEMIA PATIENTS

Although it may sound simplistic, it is important to remember that the recognition of thalassemia as a disease came about as a consequence of its effects on homozygote patients and not on the individual carriers. It is thus natural in any efforts to formulate a coherent prevention and management strategy, to concentrate efforts and resources on the patients most immediately and severely affected by the disease and work for the alleviation and elimination of its consequences.

Thalassemia families faced two acute problems in those days:

1. The improved treatment of their affected children, giving them the best chance for survival until more effective and convenient methods were developed;
2. The need to be able to have a healthy family without the fear of having another thalassemic child with each pregnancy.

A number of measures were taken to address the first problem, including the establishment of a separate clinical wing in the Cyprus Thalassemia Centre, for the treatment and follow-up of thalassemic patients; the initiation of a public education campaign to encourage blood donation, so that individual parents would not have to search themselves for blood for their children; and the free prescription of Desferal (desferrioxamine) to all thalassemia patients. These measures have had a tremendous impact on the thalassemia patients and their families over the years, greatly reducing mortality during childhood. With optimal treatment protocols, many thalassemia patients develop through puberty and adolescence without major complications. The public education on blood donation has also had a beneficial effect not only on thalassemia but on all other ailments requiring transfusion, enabling Cyprus to become self-sufficient in blood supplies. The thalassemia patients in Cyprus were thus spared from the AIDS epidemic linked to the use of imported blood, which has had a big impact on thalassemia patients in some other countries.

It was clear, however, that such a policy could not be sustained for long on its own. The existing thalassemia patients were consuming more than 50% of the available blood supplies, while more than 20% of the total drugs budget of the Ministry of Health was used for the purchase of Desferal. Furthermore, with an expected birth rate of 60-70 new patients per year, the number of patients could double in about ten years, thus stretching the limited blood supplies and other resources to the limit and compromising the quality of care not only for the existing thalassemia patients but for other patient groups as well.

In addition, although improved medical care was having a very positive effect on patient survival, treatment remained fraught with difficulties. The inadequacy of the treatment and the increasing awareness of the genetic nature of the disease, led many couples with one or more thalassemic children either to postpone pregnancies altogether, or to interrupt ongoing pregnancies without knowing the exact carrier status of the fetus. It was thus clear that neither individual families nor society at large would be able to provide the best treatment and upbringing to the existing patients, unless efforts to improve treatment were coupled to efforts to reduce new births of thalassemia patients.

3.   DEVELOPMENT OF AN EFFECTIVE PREVENTION POLICY

A number of epidemiological studies from the 1960s and early 1970s indicated that about 15% of the population of Cyprus are carriers of beta-thalassemia (Kattamis, C., et al., 1972; Modell, C.B., et al., 1972; Ashiotis, T., et al 1973; Bate, C.M., 1975), posing a major challenge in their rapid identification and proper counselling. On the basis of this frequency, about 2% of all couples were expected to both be carriers, leading to an estimate of about 3000 couples of child-bearing age. On the other hand, only about 600 patients were known on the island, with about 60-70 new patients being added every year. Most patients at that time were in the 0-10 years age range, since only the milder thalassemia intermedia cases survived past adolescence into adulthood. With such a high frequency of carriers and the large numbers of undiagnosed couples, it was considered unwise to restrict carrier identification to the families with a history of thalassemia, since such an approach would not pick up most of the at-risk families and could enhance rather than reduce the stigmatisation associated with thalassemia. It seemed quite likely that such an approach could backfire, reducing rather than increasing the voluntary participation in testing.

Indiscriminate voluntary population screening was thus considered the method of choice, but proper counselling of all persons examined presented major difficulties due to the unavailability of fully trained genetic counsellors (Modell, B., et al 1980; Mouzouras, M., et al., 1980). Preliminary voluntary screening projects on students in the third year of high school and on army recruits proved very ineffective, since the candidates did not know enough to be able to make an informed choice about the testing and there was a significant time lag between finding out about their thalassemia status and the time when they would need to use that information. Indeed it became clear from early on that the most effective way to have quick results was by addressing the section of the population at most immediate risk, i.e. new couples and the older couples with an ongoing pregnancy.

In order to attain this objective, a specialised thalassemia laboratory was established in 1978 as part of the Cyprus Thalassemia Centre, with a capacity of about 12,000 blood samples a year. Screening was offered free of charge to all requesting the service. With an estimated 6000 marriages a year, it was thus possible to pursue a policy of examining virtually all the young couples, while also gradually examining older couples, as they presented with a pregnancy.

As far as education was concerned, it was clear that it was necessary to increase general public awareness about the disease. This was pursued actively and skilfully by the Director of the Thalassemia Centre at the time, Dr. Minas Hadjiminas. Specialised information was distributed to all doctors, with an emphasis on gynaecologists, obstetricians and paediatricians, both in order to increase their awareness of the disease and to encourage people to come forward voluntarily for examination. The biology curriculum of students in the last years of high school was adapted to include the basic application of genetics with relevance to thalassemia. Public presentations were given in parallel with the effort to encourage voluntary blood donation. More general information was widely publicised in the media, which played a very important role throughout the prevention campaign.

The overall effect of this education campaign in the late 1970s and early 1980s was to raise public awareness to such a high level, that demands on the screening service far exceeded the capacity of the laboratory, thus basically being obliged to restrict the service for a number of years to the new couples and those couples with ongoing pregnancies. This resulted in the identification of about 100 new couples a year at risk for beta-thalassemia. The other important benefit of this campaign, was to reduce the demand on the genetic counselling services by individual carriers, allowing proper genetic counselling of the couples who were both carriers. Another indirect benefit from this campaign, was the impact it had on immigrant Cypriots all over the world, making them accept much more readily than other groups pre-natal diagnosis services in those countries (Modell, B. et al., 1984).

Although pre-natal diagnosis for thalassemia did not start in Cyprus until the middle of 198 1, a reduction of more than 50% was already obvious in the numbers of new births of thalassemia patients by 1980 (Angastiniotis, M.A. and Hadjiminas, M.G., 1981). This was partly due to significant numbers of couples going to London for pre-natal diagnosis from 1976 onwards, but also due to postponement of pregnancies by many couples who already had a thalassemic child, in anticipation of the pre-natal diagnosis becoming available in Cyprus. An understanding of the genetic basis of the disease may have also influenced some couples who already had one or more children with thalassemia to avoid having more children. Here it is important to emphasise one very important aspect of pre-natal diagnosis, as it has been applied to thalassemia in the late seventies (Modell, B., et al., 1982; Modell, B., 1983). The development of the techniques for pre-natal diagnosis was driven by the strong desire of couples who had already one or more children with thalassemia to have some type of choice in future pregnancies, rather than be 'forced' by the fear of having another thalassemic child to undergo an abortion, without knowing the carrier status of the fetus. Thus pre-natal diagnosis for thalassemia never evolved as a selective tool to abort thalassemic fetuses, but as a tool to rescue normal fetuses from being aborted. Despite the relatively crude nature of placentocentesis and the traumatic experiences with prostaglandin-induced abortions when compared to current techniques, their availability was highly appreciated by those in the best position to benefit from them. In fact the very first pre-natal diagnosis case that was performed in Cyprus in 1981 was on a lady who had come for a termination because she already had one thalassemic child. Needless to say that it was to the satisfaction of all of us that the fetus was a heterozygote!

4. INTRODUCTION OF THE PREMARITAL CERTIFICATE

Following the introduction of pre-natal diagnosis in Cyprus in 1981, the uptake was almost complete, leading to the birth of only 10 homozygous babies in 1981, 8 in 1982 and 8 in 1983. Thus the programme for public education and voluntary population screening and pre-natal diagnosis was already proving quite effective.

Analysis of the few cases of thalassemic babies being born indicated, however, that this was often the result of failure of some doctors to inform the couple about their risk or of laboratory errors when testing was done by private laboratories, since there was no statutory quality control system in place. At the same time, the Holy Church of Cyprus, which largely financed the construction of the Cyprus Thalassemia Centre in order to improve the treatment of the thalassemic patients, 'privately' expressed reservations on the issue of abortion following pre-natal diagnosis and encouraged ways to be sought in which the numbers of new at risk couples could be reduced. The introduction of the premarital certificate in 1983 provided the best solution to all these problems. It had a number of advantages:

1. It required all Greek-Cypriots to obtain a certificate from the Cyprus Thalassemia Centre and present it to the church authorities in order to obtain a license to get engaged or married.

2. The confidentiality of the laboratory results was in no way compromised, since the certificate presented to the Church authorities only stated that the bearer of the certificate was examined and properly advised for thalassemia. The actual diagnosis was given to individuals in the form of an identity-type card for their private use.

3. Only the certificate from the Cyprus Thalassemia Centre was accepted by the church authorities. Thus individuals who may had been examined by private laboratories with questionable quality control, had to get their results certified by the Cyprus Thalassemia Centre. Needless to say that this allowed a high degree of quality control on the work of the private laboratories and virtual elimination of laboratory misdiagnosis of cancers in subsequent years.

4. Since iron deficiency, alpha-thalassemia (Hadjiminas, M., et al., 1979; Smith, M.B. and Cauchi, M.N., 1979; Baysal, E., et al., 1995), delta-thalassemia (Triffilis, P., et al., 1991; Trifillis, P., et al., 1993) and other factors were often found to complicate unambiguous diagnosis of single individuals, the issuing of a premarital certificate to prospective couples allowed us to identify couples at greatest risk, i.e. those couples in which one or both partners had ambiguous

haematological results or those in which at least one partner had ambiguous results while the other partner was a definite heterozygote. Single individuals with diagnostic problems were presented with an 'interim' diagnosis and asked to present for re-evaluation of their results once they had found a partner. We were thus able to apply the more specialised diagnostic techniques like globin chain biosynthesis only to the individuals most at risk.

5. It was also hoped that early knowledge of the carrier status of individuals, would allow them to take this factor into account when selecting partners. Since at that time a significant number of 'arranged' marriages were taking place, especially in the countryside, this was conceived as an indirect way of reducing the numbers of new couples with thalassemia and hence the numbers of pre-natal diagnosis tests and abortions. Prostaglandin-induced abortions, sometimes as late as 21-22 weeks of gestation, were a traumatic experience for all involved and it was thus natural to reduce as far as possible the numbers of couples who had to go through this ordeal. The application of the premarital certificate thus gave the opportunity to prospective couples to be properly informed and counselled, so they would be fully prepared for the ordeals of thalassemia, in case of at-risk couples.

While there is only anecdotal evidence for any impact of the premarital certificate on choice of marriage partners, the actual numbers of couples undergoing pre-natal diagnosis per year did not show any significant reduction after the introduction of the premarital certificate. Its impact on reducing the numbers of abortions due to thalassemia was thus insignificant. On the other hand, the premarital certificate enabled the Cyprus Thalassemia Centre to place laboratory quality control on a sound footing and to ensure that all specialised laboratory testing was concentrated on the couples most at risk. With the implementation of the premarital certificate, births of new thalassemic patients dropped to very low levels, usually 0 to 2 per year since 1985.

5.   THE ERA OF MOLECULAR GENETICS

Following the discovery and use of restriction site polymorphisms in 1978 for the diagnosis of sickle cell disease, a number of other polymorphic markers were soon recognised, that enabled identification of the mutant haplotype and pre-natal diagnosis for thalassemia at about 10 weeks (Old,

J.M., et al., 1984; Wainscoat, J.S., et al., 1984; Thein, S.L., et al., 1985; Old, J.M., et al., 1990). Despite the need to collect samples from many family members for linkage analysis and the complexity of the overall approach for DNA diagnosis from chorionic villus samples (CVS), the convenience of the approach relative to second trimester diagnosis, drew increasing numbers of couples to forgo diagnosis in Cyprus at 18 weeks, in favour of CVS diagnosis in London. Thus over the years up to 1989, there was a continuous but gradual drop in the number of cases undergoing second trimester foetoscopy in Cyprus, in favour of CVS diagnosis in London.

The very success of the prevention programme for thalassemia in Cyprus, created a feeling among decision-making circles that the problem was under control, leading to decreased support to the Thalassemia Centre and relative failure to keep up with developments. Since prevention resulted in only 0-2 new patients a year, it was argued by some that the problem would gradually decrease, as the existing patients slowly died off. In view, however, of the increasing numbers of couples resorting to CVS diagnosis in foreign centres, the government eventually provided support for the establishment of the first molecular diagnostic laboratories in Cyprus in 1990. Since by that time many other genetic disorders could be diagnosed pre-natally by molecular diagnostic techniques, it was considered necessary to establish facilities that could cope with the large variety of different genetic disorders and not just with thalassemia. Thus the molecular diagnostic laboratories became part of the newly established Cyprus Institute of Neurology and Genetics.

Switching from fetuscopy or cordocentesis and globin chain separation to CVS diagnosis was completed very quickly. While only about 15% of the couples had CVS diagnosis in 1989, this was increased to about 80% in 1990 and about 91% in 1991. Nowadays only 1-2% of couples undergo second trimester diagnosis for various reasons, while the uptake of pre-natal diagnosis is about 98% (Cao, A., 1987; Angastinotis, M., et al., 1988; Angastinotis, M., 1992). It may thus be justified in the minds of some people to think that the current approach at containing the problem of thalassemia is a viable long term approach: but is it?

6.   WHICH WAY FORWARD?

Despite the greater convenience and acceptability of CVS diagnosis over second trimester diagnosis, there is also increasing awareness of the value of human life during fetal development and hence a search by many couples for more acceptable means of prevention and management of thalassemia. Preimplantation diagnosis may provide the answer to some couples, but it is unlikely that such a technically demanding procedure will become as widely applicable as CVS diagnosis.

Furthermore, although pre-natal diagnosis was well accepted in Cyprus as an immediate solution to an acute problem, it is much less acceptable as a solution to many other populations for a variety of social or religious reasons. In addition, the logistics of the prevention programme in Cyprus make it seem very unlikely that the same approach can be easily applied to countries with much larger populations and a lower incidence of thalassemia. It is instructive to remember that despite community-directed approaches in the prevention of thalassemia in Britain, the success rate is only about 50% after 20 years of prevention, and that is mostly due to the high uptake of the services by the immigrant Cypriot population (Modell, B., et al., 1984). Thus although efforts to promote prevention in other countries should no doubt continue (Angastinotis, M. et al., 1995), it should be appreciated from the start that the success of Cyprus and a handful of other countries or regions may not be so easy to repeat, nor should it be the main focus of new prevention programmes.

In face of the gradual improvement of the clinical management of thalassemia patients in most countries and the limited application of effective prevention to a handful of countries, it is reasonable to expect that the numbers of thalassemic patients would increase rapidly over the next 20 years in most countries. Even in Cyprus, there is increasing resistance by at-risk couples to accept pre-natal diagnosis and possible abortion as their only option for the future.

On strictly objective criteria, it is also unacceptable to consider abortion as a viable long term option in the prevention of thalassemia and other genetic disorders. Every individual human being has a unique combination of about three billion bases in its genetic code, yet the decision for abortion after pre-natal diagnosis is based in most genetic disorders on the change of one or another base, without any regard to the potential encoded by the rest of the genome. It may be sometimes that other genes can complement the missing function, but a diagnosis based on detecting specific molecular changes will most often fail to detect such possibilities, leading to abortion of fetuses that would otherwise have had prospects for a good quality of life. This has certainly been the case with thalassemia intermedia in Cyprus (Weatherall, D.J. et al., 1981; Wainscoat, J.S. et al., 1983a; Wainscoat, J.S. et al., 1983b), despite efforts to take due account of the existence of quite mild forms of the disease in some patients. In any case it seems most sensible to work in developing approaches that will relieve the individual of the genetic burden of the genetic defect, giving an opportunity for the unique genetic combination to produce a human being with a good quality of life, rather than discarding the whole individual.

It should thus be clear that a viable long term prevention programme for a genetic disease like thalassemia must have the existing patients at its focus and not simply the prevention of the birth of new patients. As in all other areas of medicine and science, the objective must be to develop ways to ameliorate the consequences of the genetic defect on the overall quality of life of the patients and their families. In some genetic diseases the problem may be managed satisfactorily by dietary modifications, in others by enzyme replacement or bone marrow transplantation; it is also our hope that one or more of the large array of gene therapy approaches that are being developed will provide the best answer for thalassemia and other genetic disorders. Indeed this is becoming increasingly the focus of work in Cyprus, as it is already the focus of the work of so many other groups around the world. Undoubtedly, however, the development of such approaches will not only lead to the effective and acceptable treatment of existing patients in all countries, but will also transform pre-natal diagnosis from a tool that rescues normal fetuses but often leads to the abortion of genetically affected fetuses, into a tool for the diagnosis and effective therapy of affected fetuses as early as possible.

ACKNOWLEDGEMENTS

The work described here obviously does not represent the work of myself alone, or of any other single individual, but it is a brief review of the collective work of numerous individuals involved in the Cyprus prevention programme for thalassemia over the last twenty years. I would like, however, to acknowledge in particular the contributions of all my colleagues at the Cyprus Thalassemia Centre and especially those of its former director, Dr. Minas Hadjiminas and its current director, Dr. Michael Angastiniotis. I would also like to acknowledge the contributions of Dr. Lefkos Middleton, Director of the Cyprus Institute of Neurology and Genetics and of all the staff of the Thalassemia Molecular Genetics Group at the Institute.

REFERENCES

Angastiniotis, M.A., Hadjiminas, M.G., 1981, Prevention of thalassemia in Cyprus, *Lancet*, 1, 8216, pp. 369-371.

Angastiniotis, M., Kyriakidou, S., Hadjiminas, M., 1988, The Cyprus Thalassemia Control Program, *Birth Defects*, 23, 5B, pp. 417-432.

Angastiniotis, M., 1992, Management of thalassemia in Cyprus, *Birth* Defects, 28, 3, pp. 38-43.

Angastiniotis, M., Modell, B., Englezos, P., Boulyjenkov, V., 1995, Prevention and control of haemoglobinopathies, *Bulletin of the World Health Organisation*, 73, 3, pp. 375-386.

Angel, J.L., 1996, Porotic hyperostosis, anemias, inalarias, and marshes in the prehistoric Eastern Mediterranean, *Science,* 153, pp. 760-763.

Ashiotis, T., Zachariadis, Z., Sofroniadou, K., Loukopoulos, D., Stamatoyannopoulos, G., 1973, Thalassemia in Cyprus, *British Medical Journal*, 2, 857, pp. 38-42.

Bate, C.M., 1975, Thalassemia in Cyprus, *Proc R Soc Med*, 68, pp. 514-516.

Baysal, E., Kleanthous, M., Bozkurt, G., Kyrri, A., Kalogirou, E., Angastiniotis, M. Ioannou, P., Huisman, T.H., 1995, Alpha-thalassemia in the population of Cyprus, *British Journal of Haematology*, 89, pp. 496-499.

Cao, A., 1987, Results of programmes for ante-natal detection of thalassemia in reducing the incidence of the disorder, *Blood Review*, 1, pp. 169-176.

Hadjiminas, M., Zachariadis, Z., Stamatoyannopoulos, G., 1979, Alpha-thalassemia in Cyprus, *Journal of Medical Genetics,* 16, pp. 363-365.

Kattamis, C., Haidas, S., Metaxotou-Mavromati, A., Matsaniotis, N., 1972, Beta-thalassemia, G-6-PI3 deficiency, and atypical cholinesterase in Cyprus, *British Medical Journal*, 3, 824, pp. 470-471.

Modell, C.B., Benson, A., Wright, C.R., 1972, Incidence of beta-thalassemia trait among Cypriots in London, *British Medical Journal*, 3, 829, pp. 737-738.

Modell, B., Mouzouras, M., Camba, L., Ward, R.H., Fairweather, D.V., 1980, Population screening for carrier of recessively inherited disorders, *Lancet*, 2, 8198, p. 806.

Modell, B., Mouzouras, M., 1982, Social consequences of introducing ante-natal diagnosis for thalassemia, *Birth Defects,* 18, pp. 285-291.

Modell, B., 1983, Screening for carriers of recessive disease, *Proc. Annu. Symp. Eugen. Soc.*, 19, pp. 139-147.

Modell, B., Petrou, M., Ward, R.H., Fairweather, D.V., Rodeck, C., Varnavides, L.A., White, J.M., 1984, Effect of fetal diagnostic testing on birth-rate of thalassemia major in Britain, *Lancet*, 2, 8416, pp. 1383-1386.

Mouzouras, M., Camba, L., Ioannou, P., Modell, B., Constantinides, P., Gale, P., 1980, Thalassemia model of recessive genetic disease in the community, *Lancet,* 2, 8194, pp. 574-578.

Old, J.M., Petrou, M., Modell, B., Weatherall, D.J., 1984, Feasibility of ante-natal diagnosis of beta thalassemia by DNA polymorphisms in Asian Indian and Cypriot Populations, *British Journal of Haematology*, 57, pp. 255-263.

Old, J.M., Varawalla, N.Y., Weatherall, D.J., 1990, Rapid detection and pre-natal diagnosis of beta-thalassemia: studies in Indian and Cypriot populations in the UK, *Lancet*, 336, 8719, pp. 834-837.

Smith, M.B., Cauchi, M.N., 1979, Quantitative studies of Hb Bart's levels and red cell indices in alpha thalassemia trait in Mediterraneans, *Pathology*, 11, pp. 621-627.

Thein, S.L., Wainscoat, J.S., Old, J.M., Sampietro, M., Fiorelli, G., Wallace, R.B., Weatherall, D.J., 1985, Feasibility of pre-natal diagnosis of beta-thalassemia with synthetic DNA probes in two Mediterranean populations, *Lancet*, 2, 8451, pp. 345-347.

Triffilis, P. Ioannou, P. Schwartz, E., Surrey, S., 1991, Identification of four novel delta-globin gene mutations in Greek Cypriots using polymerase chain reaction and automated fluorescence-based DNA sequence analysis, *Blood,* 78, pp. 3298-3305.

Triffilis, P., Kyrri, A., Kalogirou, E., Kokkofitou, A., Ioannou, P., Schwartz, E., Surrey, S., 1993, Analysis of delta-globin gene mutations in Greek Cypriots, *Blood*, 82, pp. 1647-1651.

Weatherall, D.J., Pressley, L. Wood, W.G. Higgs, D.R., Clegg, J.B., 1981, Molecular basis for mild forms of homozygous beta-thalassemia, *Lancet*, 1, 8219, pp. 527-529.

Wainscoat, J.S., Old, J.M., Weatherall, D.J., Orkin, S.H., 1983a, The molecular basis for the clinical diversity of beta thalassemia in Cyprus, *Lancet*, 1, 8336, pp. 1235-1237.

Wainscoat, J.S., Kanavakis, E., Wood, W.G., Letsky, E.A., Huehns, E.R. Marsh, G.W., Higgs, D.R., Clegg, J.B., Weatherall, D. J., 1983b, Thalassemia intermedia in Cyprus: the interaction of alpha and beta thalassemia, *British Journal of Haematology*, 53, pp. 411-416.

Wainscoat, J.S., Old, J.M., Thein, S.L., Weatherall, D.J., 1984, A new DNA polymorphism for pre-natal diagnosis of beta-thalassemia in Mediterranean populations, *Lancet*, 2, 8415, pp. 1299-1301.

Chapter 5

Some developments in genetic screening in Finland

VEIKKO LAUNIS
Department of Philosophy
University of Turku
Finland

1.   INTRODUCTION

For centuries, Finland has been politically and geographically isolated, and the genetic history of the current population is relatively well known. Since 1965, over 30 rare genetic diseases have been found which are markedly enriched in the Finnish population and which beautifully demonstrate the 'genetic founder' effect. The population history suggests that scattered isolated populations with from under one hundred to a few hundred inhabitants were formed several hundred years ago, and that these subpopulations remained relatively separate from other subpopulations until recent years. The result was close inbreeding, which enriched the founder genes and led to the not uncommon occurrence of some single-gene diseases. This group of diseases was named the 'Finnish disease heritage', which for many years was the main stimulus for genetic studies and plans for screening in Finland. (de la Chapelle, 1993).

Meanwhile, several diseases which are prominent in other European countries occur in Finland only rarely. Examples of such extremely uncommon or totally absent single-gene diseases are Phenylketonuria, Huntington's disease, cystic fibrosis, galactosemia and albinism. Because of these losses, some of the diseases which have been particularly suitable for screening and genetic screening in other western countries are unsuitable for screening in Finland.

R. Chadwick et al. (eds.), The Ethics of Genetic Screening, 69–74.

Finland has, however, received its share of the genetic load of the common multifactorial diseases as well. Finland has been at the top of the international list in coronary heart disease mortality, due evidently in part to the exceptional prominence in the population of the apolipoprotein E4 and the low frequency of apoliprotein E2; these findings explain part of the high serum cholesterol values of the Finns. Insulin-dependent diabetes mellitus is another multifactorial disease which is more common in Finland than in any other country. The incidence in children is in fact about four-fold compared to most Central European countries, and about 40-fold compared to the incidence in Korea, Japan, and mainland China.

For two years in the late 1960s, phenylketonuria screening was performed using the Guthrie test in all newborns for two years but no new cases were found. This led to long-lasting pessimism regarding screening, and projects were postponed for years. At present, several genetically based testing programmes have been active for years, but true genetic screening of newborns or of geographically selected populations or carrier families is only in its initial stages.

2.   ONGOING SCREENING PROGRAMMES

2.1   Screening of pregnant women older than 35

The programme, started in 1979, has now been extended over the whole country. The purpose is to detect Down's syndrome and other fetal chromosome disorders. Samples are taken from placenta (chorion villus biopsy) or amniotic fluid (amnioncentesis). The test is voluntary; it is offered at maternity clinics, and is performed at university hospitals which also provide genetic counselling.

2.2   Screening of all pregnant women

This programme originally covered only the Eastern part of Finland but has gradually expanded to almost the whole country. The aim of the test is to detect fetuses with Down's·syndrome, neural tube defect or congenital nephrotic syndrome. Alphafetoprotein (AFP) and human chorionic gonadotropin (HCG) are measured first in the mother's serum, then, if abnormal, in an amniotic fluid sample. The test is voluntary, and those at risk are offered an opportunity for further examinations.

2.3   Screening of selected families

The aim is to detect persons at risk for familial colonic cancer (there are about 70 Finnish families that are carriers of this disease) or persons at risk for fragile X syndrome.

3.   FUTURE GENETIC SCREENING PROGRAMMES

3.1   Neo-natal screening for genetic susceptibility to diabetes.

The aim is to detect genetic susceptibility to diabetes mellitus. The programme has started in the hospital administration districts of Turku and Oulu, and the plan is for expansion to the whole country by the end of 1996. The informed consent of the family is required. Psychological aspects of the screening will be carefully studied to explore the way in which the families have experienced the screening and their emotional reactions.

3.2   Screening of mothers (and husbands of carriers)

There are plans for the testing of approximately 2000 Finnish women for AGU (aspartylglucosaminuria) disease, which causes mental retardation. A psychological examination will be carried out both after the test and after the birth of a child. Genetic counselling will be available for all carriers.

There are some hereditary diseases which occur in Finland but are unknown in the rest of Europe. These include for instance Salla disease and AGU (aspartylglucosaminuria) disease (causing mental retardation); HOGA

disease (gyrate atrophy, causing blindness at the age of 20-40); and congenital nephrotic syndrome. It is possible that screening programmes for
these diseases will be introduced in the near future. (Some tests for AGU
(aspartylglucosaminuria) disease have already been carried out.)

4.   THE STATE OF THE PUBLIC DEBATE ON
         GENETIC SCREENING

Up to the present, there has been relatively little public discussion on the
ethical implications of genetic screening in Finland. The following ethical
issues have been discussed to some extent.

4.1   Informed consent

Despite the voluntary nature of the screening tests, there has been
concern as to whether the consent given by persons taking a test is adequately informed. One demand has been that the ability of those involved to
understand the relevant technical aspects of genetics should be developed in
order to ensure the possibility of adequate genetic counselling (Norio, 1991;
Norio, 1997). There has also been concern over the possibility of less
voluntary - or even obligatory - forms of genetic testing in the future, for
instance in job hiring (Simonsuuri-Sorsa, 1991; Hietala et al., 1995;
Simonsuuri-Sorsa, 1997).

4.2   Genetic testing in insurance

A study has recently been carried out in Finland dealing with attitudes
towards genetic testing among the general population and families (as well
as other relatives) of patients with a genetic disease. The findings of the
study indicate that Finns have a relatively favorable attitude toward genetic
screening. Most of the respondents thought, for instance, that screening tests
should be available for anyone wanting to know more about his or her own
genes. Furthermore, Finns seem to be very confident that they will retain
their individual freedom both to participate in genetic tests and to decide

how the test results will be applied. The worries expressed in the study concerned chiefly the possibility that unauthorised groups might obtain information from tests or that test results might lead to social discrimination - especially in insurance policies. (Hietala et al., 1995; see also Kosonen, 1982).

It seems that at least at present the Finnish insurers consider genetic testing in the risk assessment of insurances an unattractive instrument. Several reasons can be found for this attitude. First, the insurers wish to avoid unnecessary social discrimination as long as it is possible. Second, the diseases most often discussed in this context (i.e. Huntington's disease and cystic fibrosis, for which specific tests are already in use) are particularly rare in Finland. Third, the specificity of the test in the common polygenic disorders (such as malignant and cardiovascular diseases) is so inaccurate that the cost-benefit would probably be very poor. (Palotie & Pelkonen 1994).

However, there is no reason to believe that the insurers will not ask applicants to take genetic tests (or to show test results) in the future if new and better techniques are available and the use of them in the risk assessment is considered economically reasonable.

ACKNOWLEDGEMENTS

I am indebted to Professor Pertti Aula, Professor Risto Pelkonen and Professor Olli Simell for helpful comments and technical information and to Mrs Ellen Valle for improving the style of this paper.

REFERENCES

de la Chapelle, A., 1993. 'Disease gene mapping in isolated human populations, The example of Finland', *Journal of Medical Genetics,* 30, pp. 857-865.

Hietala, M. et al., 1995. 'Attitudes toward genetic testing among the general population and relatives of patients with a severe genetic disease, A survey from Finland', *American Journal of Human Genetics,* 56, pp. 1493-1500.

Kosonen, T., 1982. 'Salassapitovelvollisuus terveydenhuollossa' [Confidentiality in health care], in K. Achté et al. (eds), *Lääkintäetiikka [Medical ethics],* Suomen Lääkäriliitto, Vaasa, pp. 84-114.

Norio, R., 1991. 'Lääketieteellisen genetiikan eettiset ongelmat' [Ethical issues in medical genetics], in P. Löppönen, P. H. Mäkelä and K. Paunio (eds), *Tiede ja etiikka [Science and ethics]*, WSOY, Porvoo, pp. 328-342.

Norio, R., 1997. 'Perinnöllisyysneuvonta ja sikiödiagnostiikka' [Genetic counselling and fetal diagnosis], in V. Launis and J. Räikkä (eds), *Geenit ja etiikka [Genes and ethics]*, Edita, Helsinki, pp. 81-91.

Palotie, L. and Pelkonen, R., 1994. 'DNA-analyysit ja henkilövakuutus' [DNA-analysis and Personal Insurance], *Duodecim*, 110, pp. 537-539.

Simonsuuri-Sorsa, M., 1991. 'Työympäristön riskit ja tutkimusetiikka' [The risks in the working environment and research ethics], in P. Löppönen, P. H. Mäkelä and K. Paunio (eds), *Tiede ja etiikka [Science and ethics]*, WSOY, Porvoo, pp. 358-368.

Simonsuuri-Sorsa, M., 1997. 'Työntekijöihin kohdistuva geneettinen testaus ja eettiset ongelmat' [Genetic testing of employees, Ethical issues], in V. Launis and J. Räikkä (eds), *Geenit ja etiikka [Genes and ethics]*, Edita, Helsinki, pp. 92-107.

Chapter 6

Genetic screening: Ethical debates and regulatory systems in France

ANDRÉ BOUÉ
Comité Consultatif National d'Ethique pour les Sciences de la vie et de la Santé
Paris
France

1. THE ETHICAL DEBATES

Genetic screening and more generally predictive medicine have been the subjects of debate in different French regulatory bodies.

The National Ethics Committee, Comité Consultatif National d'Ethique pour les Sciences de la Vie et de la Santé (CCNE), was created by a governmental decree in November 1983, and in May 1985 it published a report and recommendations on pre-natal and perinatal diagnosis. The CCNE has subsequently published several reports and recommendations: in December 1989, on the use of DNA fingerprints; in June 1991, on the application of genetic tests for individual, familial or population studies; and in June 1993 on the use of maternal biological markers for pre-natal diagnosis of trisomy 21.

In May 1988, the Government established the National Commission for Medicine and Biology Reproduction and for Pre-natal Diagnosis. Pre-natal diagnosis analysis may only be carried out at public or private laboratories authorised by the Commission.

In 1988 there began a large debate concerning the requirement for laws regulating different aspects of research and medical activities such as assays in human beings, transplantations, medically assisted procreation, pre-natal

R. Chadwick et al. (eds.), The Ethics of Genetic Screening, 75–80.

diagnosis, and predictive medicine including genetic tests. Different regulatory bodies have contributed to the debate on genetic screening.

The Council of States (Conseil d'Etat) published a large report and proposals for laws in April 1989. The Government, the National Assembly, and the Senate had working groups on bioethics in view of the preparation of laws which were first adopted by the National Assembly in December 1992. These laws were then discussed several times between the National Assembly and the Senate and were definitively adopted in July 1994 with a large majority and a consensus across the political parties.

1.1   The laws

The legislation (*Journel Officiel de la République*, 1994) concerning pre-natal diagnosis requires genetic counselling, and creates reference clinical and biological centres in some authorised public or non profit private hospitals. The indication for the termination of a pregnancy after pre-natal diagnosis must be submitted to practitioners belonging to these reference centres. A few of these centres may be authorised to perform pre-implantation diagnosis when a couple has a high risk of giving birth to a child suffering from a severe genetic disease clearly identified in the parents.

A chapter of the law relates to 'Predictive medicine': genetic testing or screening can only be carried out for medical indications or scientific research and with the written consent of the subject. A decree (not yet published) will detail the conditions for the application of this chapter of the law. An interesting aspect of this law is that it will be again submitted to the National Assembly and the Senate after five years (in 1999) in order to modify some articles.

1.2   The recommendations of the National Ethics Committee

In October 1995, the CCNE (Comité Consultatif National d'Éthique pour les sciences de la vie et de la santé) published a large report on genetic screening (Comité Consultatif National d'Éthique pour les sciences de la vie et de la santé, 1997, pp. 188). Its recommendations included the following :

1. An examination of the genetic characteristics of an individual may, whatever the result, have profound repercussions on the life of the

person who submits to it. For his independence to be respected, he must have as complete an understanding as possible of the consequences of his decision to accept the test or not. Such understanding implies information on the nature of the test, the significance of results, the possible existence of prevention and therapy and consequent constraints. This information must be imparted by a professional person with good knowledge of medical genetics, must be direct and oral so that a dialogue can take place, and must then be put down in written form.

Any determination of the characteristics of the genotype of an individual must only be undertaken for medical purposes by prescription or for scientific purposes and only if the subject has specifically given written consent.

The results of the tests must be communicated in person by a physician whose competence permits a full explanation of the significance of the results. A follow-up of the patient must be provided in order to alleviate possible psychological repercussions because of the results of the test, be they positive or negative.

Certain kinds of information may have potentially harmful effects on the individual. He may therefore refuse to be given the results of the test and his right not to know must always be respected.

2. Medical confidentiality must be observed, and information should not be disclosed without consent to third parties, even other family members. If discovery of a genetic abnormality has consequences for family members perhaps justifying the offer of testing, then other members of that family must be approached by the requesting individual and not by the physician. If the patient refuses to alert members of his family of a risk revealed by the genetic examination, the physician cannot warn them of the possible risk of developing a disease or of transmitting it to their descendants. The physician must inform the index patient of the responsibility he is incurring and do his best to convince him that he should inform his relatives. If that procedure fails, the principles of medical confidentiality and of the duty to inform patients and their families of a risk which may be averted by preventive measures, will be in contradiction. The physician will be confronted with a serious ethical conflict on which society must form a view, taking into account the unacceptability of refusal to help endangered persons, particularly if children are involved.

The study of genetic characteristics of children should not be systematic. It must always be related to a specific case and based on an analysis of medical and familial data. Parents and attendant physician must only request a test for a child if the disease associated with his genotype may become manifest before the age of 18 or if preventive therapy before 18 years of age may be of benefit. A child who has been genetically tested and for whom medical follow-up is required, must be informed as soon as he is able to understand the procedures. In cases where the test would lead to an appraisal of a risk for the child's future descendants, his family's duty is to inform the child as soon as he or she reaches reproductive age and he is able to understand and decide for himself to submit to testing. Transmission of genetic information from one generation to the next may be necessary. Provision must therefore be made for the conservation of a family's genetic data for at least a generation and also for at-risk individuals to be given this information when it becomes useful to do so.

3. Computer storage of identifiable data relating to persons whose samples have been tested must be carried out in such a way that confidentiality is protected in observance of legal rules and of recommendations issued by the CCNE in its previous opinions.

Biological samples which may be required at a later date for further testing, for purposes of diagnosis or verification, must be preserved so as to be able to respond to a patient's needs.

Should there be an extension of research to a domain not foreseen at the time of sampling, consent must again be obtained.

When the collection of biological samples is made in the framework of a research programme, the instigators of the research are duty bound to complete such research work with the means available and in the conditions described at the time of consent being given by the individuals who have been sampled.

Non use for a prolonged period of time of such collections by scientists who are making no progress in their research could be harmful to the legitimate expectations of those persons who have consented to use being made of samples of their DNA. It would therefore be necessary to set reasonable deadlines after which access to the collections would be open to other scientists wishing to work on the protocol for which consent was given. In cases where the investigators abandon the research project, they should inform persons concerned of consequent modifications.

4. Use of the results of a study of genetic characteristics for purposes other than medical or scientific, for instance for an insurance contract or for employment, is prohibited even though the tests may have been requested by the persons concerned or with their consent.

Instances of a study of genetic characteristics being useful for preventing work-related diseases are rare indeed in the present state of scientific knowledge. The use of genetic testing in occupational medicine must therefore be exceptional and rigorously restricted to cases on a limited list for which the risk for the individual is sufficiently established and cannot be removed by changes to the work environment.

5. Approval procedures by the Drugs Agency (Agence du Médicament) must be set up for the reagents used in genetic testing protocols. Their sale must be regulated and so must approval and supervision of laboratories performing the tests.

   When studies are practised on a large number of individuals, precise preliminary work must be done to evaluate the predictive value of the tests and the usefulness of preventive and curative measures to be recommended to people selected by the tests. This must be done before authorisation is granted for the studies.

   Evaluation should not be based solely on medical criteria. It must also take into account the various aspects of quality of life and the way in which it will be modified either when screening is performed or because of constraints due to prevention. The evaluation of a genetic screening and prevention programme must include the fact that such a programme can only be effective if the protocol is considered acceptable by the target population and by the medical profession.

6. The attitudes of individuals and their families to genetic screening and preventive measures are a consequence of the quality of the medical information given to those concerned. It is essential, therefore, that members of the medical and para-medical professions should attend appropriate medical genetics training courses at universities and that practising physicians be provided with refresher courses.

   Pedagogical information imparted during secondary education, in biology or philosophy classes, should make it possible to reduce the risk of discrimination or stigmatisation due to knowledge of genetic characteristics.

   Associations representing families with an interest in a genetic disease should be encouraged in their efforts to widen medical and scientific information.

   A close watch must be kept on the quality of the media information written for the general public which may lead to false hopes because of a penchant for sensational news. Another possible danger is that the sizeable potential market for genetic screening may lead to vested financial interests which could be detrimental to the truthfulness and independence of information.

REFERENCES

Comité Consultatif National d'Éthique pour les sciences de la vie et de la santé, 1997.
Génétique et Médecine, de la prédiction à la prévention, La Documentation Française, pp. 188.
Journal Officiel de la République Française, 1994. pp. 11060 - 11068.

Chapter 7

Screening in Germany: Carrier screening, pre-natal care and other screening projects

TRAUTE SCHROEDER-KURTH
University of Wurzburg
Germany

1. OFFICIAL BODIES' OPINION

In Germany no carrier screening projects, no pilot studies about the requirements for population screening or its acceptance in a population are presently financed by institutions like the German Research Foundation (Deutsche Forschungsgemeinschaft, DFG) or Ministries of the Republic or the States. No proposal for a carrier screening programme or a pilot study has been submitted to the DFG. One project proposal for cystic fibrosis (CF) carrier screening has been rejected by this institution in 1989 because the outline did not meet the conditions for research-funding.

In 1994 the Senate of the DFG established a commission to advise the Senate on rules of conduct in connection with ethical problems evoked by research and applied gene-technologies. This commission held a hearing about the advantages and problems of population screening in general. The commission decided to observe carefully further developments, but found no reason for any recommendations to the Senate. This means that screening projects can be submitted and would be supported if they fulfil the requirements. The German Parliament intended to call a National Bioethical Commission to advise the government in 1994, however, this was not done. The Ministry of Health formed an Ethics Council in 1995, which works

R. Chadwick et al. (eds.), The Ethics of Genetic Screening, 81–87.
© 1999 *Kluwer Academic Publishers. Printed in the Netherlands.*

together with the Ministry of Justice and, at present with a special working unit, on new reproduction technologies. Screening is not on the agenda.

The Federal Medical Association (Bundesärztekammer) also established an independent 'Central Ethical Commission', which would answer basic questions if population screening was proposed. There was no reason to inform the Research Council of the Medical Association to change their opinion about 'Genetic Screening' published in 1992. This 'Memorandum' (Bundesärztekammer, 1992) dealt exclusively with population carrier screening, i.e. in adults for genetic defects without direct medical indication like molecular genetic testing for cystic fibrosis carriers. It made clear, that pre-natal screening for chromosome abnormalities in older women on their own request is different, although both types of screening are considered in many respects of similar nature and connected with similar problems. The main question remained unanswered: which goal would be persued with population screening, what kind of health education and public information must precede, would individual counselling be a necessary prerequisite and if so, who should be in charge of counselling and who pays for everything?

The professional German Medical Genetics Association (Berufsverband Medizinische Genetik) published a 'Statement about Screening of Heterozygotes' in 1990 (p. 6-7), pointing to the danger of discrimination of carriers within an uneducated society.

A working group was established by the Federal Ministry of Research and Technology (Bundesministerium für Forschung und Technologie) in 1989 in order to evaluate the 'Ethical and Social Aspects of the Human Genome Analysis'. Their results were published in 1990, including a distinct chapter on population screening and its problems, the screening with versus without medical indication and the necessity of informed consent versus informed refusal.

2.   SCREENING IN PRACTICE

Actually, in hospitals and genetic ambulances or in the practice of gynaecologists it daily becomes apparent that 'screening' in the sense of search for diseases, dispositions, for carriers within families with index patients or signals for certain risks has almost become a standard of medical care. There has been a mother-child passport in place in Germany for many years, which means thorough investigations and observations of the mother's health during pregnancy plus presently three ultrasound examinations of the developing fetus with special care if unusual developments are found. This is a screening program of the whole population of pregnant women, aimed at the safeguard of mother and child until birth. Screening of newborns for treatable diseases was introduced in 1976. Currently, about 96% of the newborn babies in Germany are investigated for phenylketonuria, galactosemia and hypothyroidism (Schmidtke, 1996). One laboratory in Berlin is screening the new-born population of this city for the Biotinidase defect (pers. communication, Cobet 1994).

Voluntary 'screening' in this sense, following a medical indication, also applies to maternal serum testing for alpha-feto-protein (AFP), human choriongonadotropin (HCG) and estriol - the so-called Triple-test - for women who have a medical indication for an invasive pre-natal diagnostic procedure, but are ambivalent or are particularly cautious. Other women above 27 years of age use the test to get an indication for pre-natal diagnosis. They all have to learn during a pre-test counselling, how to interpret the results of the test in order to include the risk figure in their decision making.

The indication for invasive pre-natal diagnosis is still related to an increased risk, i.e. standard care starts for women above 35 years of age in case of risk for chromosomal abnormalities. The last complete analysis of the numbers of users showed that 53% of pregnant women 35 years and older received pre-natal chromosome analysis after amniocentesis or chorion villi biopsy in 1987 (Schroeder-Kurth, 1989). Other medical indications are related to higher risk for specific diseases. Pre-natal diagnosis for chromosome abnormalities in cases of severe anxieties are possible and are covered by the obligate as well as by most private insurance if there is a doctor recommendations.

Both, ultrasound and pre-natal diagnosis can be refused by women without suffering disadvantages in medical care or insurance. The actual disadvantage, however, would emerge around birth and, in particular, in case of a complicated birth with a baby who urgently might need treatment of an expert.

True 'carrier-screening' for example for CF-heterozygotes as part of standard care for pregnant women has not been announced by the professional Medical Board (Beirat der Bundesärztekammer). That means, that today no 'screening' for genetic defects takes place unless a couple has a problem or a pregnant woman asks for a specific test. The information and education by the mass media about screening possibilities is rather rare. If there is coverage in a journal for parents or in a television programme, then more questions will be asked during pre-testing counselling, however, normally there is no request for a specific screening.

3.   PRIVATELY FUNDED SCREENING PROGRAMMES

There have been a few exceptions from the general abstain from screening without medical indication which should be noted.

In 1992 one Institute of Human Genetics reported at the Annual Meeting of the Society of Human Genetics at Würzburg about a carrier screening for CF, which was offered during counselling before pre-natal diagnosis for other reasons, irrespective of a family history with the disease (Issue of Conference). The counsellors tried various ways to inform the pregnant women:
1.   by direct verbal information during counselling,
2.   by asking before counselling if they would like to get information, and
3.   by leaflet information and a questionnaire.

The results of the trial showed differences in acceptance of the test:
1.   during counselling: 22% decided to be tested,
2.   when asking if they would like to get the information: 12.3% accepted, and
3.   after leaflet and questionnaire: 16.6% wanted the test.

There was one pilot study of Coutelle's group from Berlin-Buch which reported the results of CF-carrier testing of 638 pregnant women from 1989-1993, (Gille et al., 1991). This paper was criticised by Schmidtke, who pointed to basic differences in the approaches and attitudes in the pre-test counselling leading to almost unanimous consent to the screening offer. (Schmidtke, 1994). Coutelle answered immediately and explained his general line of argument - full and detailed discussion with the patient only after a 'positive' result. (Coutelle, 1994). The objection to this pilot study thus aimed at the 'informed consent', not against the technical set-up of the test.

4.   SCIENTIFIC 'SCREENING'

Other 'screening' programmes have been reported for example in the abstracts of the 6th. Annual Meeting of the Society of Human Genetics at

Düsseldorf, in 1994 (Conference Proceedings). The term 'screening' was quoted 24 times, mostly in other connections than 'population screening'. Two interesting contributions came from I Nippert. She studied the views, interests and the concerns of so-called 'key-persons in health care' in 4 countries and demonstrated disagreements about the desirability of population - based CF-carrier screening because of the potential magnitude of providing CF-carrier screening to the whole population and its unknown social implication. On an inter and intra country level those who were against CF-carrier screening (such as the representative of the CF Self-help Group Kruip, 1993), opposed it either because of the abortion issue related to it or because of the unassessed potential of negative implications, i.e. the eugenic goal and effect. Proponents stressed the potential of providing information for allowing autonomous reproductive decisions, giving people the option of avoiding the birth of an affected child. J. Schmidtke concentrated on the problem of autonomy and genetic screening. He explained two basically different ways for deciding who should make the final decision for or against testing: the individual or third parties through the health system.

At the 9th Annual Meeting of the Society of Human Genetics, held at Innsbruck from 16 to 19 April 1997, there were about 20 abstracts using the term 'screening', mostly for mutation analysis or search for certain candidate genes among patients and their relatives (Conference Proceedings, Medizinische Genetik, 1997). All of these projects are scientific studies, none is an applied test to serve a distinct population at risk or the general population. J. Schmidtke again presents a paper on population screening and its general considerations.

In 1996, Schroeder-Kurth, Lunshof and Schäfer finished a three year study of the development of new technologies in human genetics and its consequences. The results were published under the heading. 'Human Genetics and Society'. One part of the project dealt with 'population screening'. Representatives of patient-organisations and geneticists were asked to answer questions about the understanding and definition of the term 'screening', its meaning and its social implications, in particular, concerning pre-natal care and diagnosis, the value of a 'medical indication' and the Triple-test with its specific problems. One result out of this series of questions and answers showed a clear decision for a medical indication for women at risk by the patient-group. (Schroeder-Kurth et al., 1996). This contrasts the opinion of Schmidtke and Wolff (1991) and Schmidtke (1995) who supported solely the autonomy of women irrespective of risk, which would mean surrender of the medical indication. Other human geneticists disagreed for various reasons. (see discussion in Medizinische Genetik, 1991). The importance of precounselling and informed consent was stressed. Screening for heterozygosity of severe diseases seemed acceptable for 53% of the geneticists and 47% of the affected. In conclusion, it became again evident, that the feasibility of new methods for direct diagnosis for pre-and postnatal screening deserved serious consideration before implementation. Until now there has been no sign, however, that the Federal Republic of Germany will be pushed into such development. The concept, however, of the 'medical indication' has come again into focus recently with the attempt of the Government to reorganise the Health System.

5.   CONCLUSIONS

The fundamental questions remain unanswered. What is the goal of screening of heterozygotes in general? What is the goal of screening pregnant women and the unborn? Who benefits or should benefit from the result - the individual or the society? How would it be possible to serve the population well so that the individual still can decide for or against the test without disadvantage? Women organisations oppose the pregnancy screening because they feel an enormous social pressure towards the technical control of their pregnancy. Would it be the same if carrier screening is offered on a voluntary basis? How would it be possible to serve the society well so that the society in general is satisfied with any result, as long as the decision remains personal: may it be information about the genetic status only, may it be for personal orientation like professional match-making for marriage, for counselling, or for the decision, not to have children with a carrier partner, or for family planning including pre-natal diagnosis and the so-called therapeutical abortion. (Pander et al., 1992; Schmidtke and Vogel, 1990; Schroeder-Kurth, 1996).

Carrier-screening seems to be the crucial test as to whether it will be possible to offer screening based on an individual decision to use or not to use a test, will avoiding any pressure for decision making in certain directions by third parties. Today, however, it seems unreasonable to demand public support or reimbursement of health insurance for such screening programs. It would not have top priority in a country with great problems in financing the treatment of the sick. The other possibility would be to allow testing via the private sector and leave the development of this 'experiment' in our society to the free market and testing to commercialisation. Then genetic data would be available to those who want to be informed and are able to pay for the test, but proper counselling and interpretation would be missing; in addition commercialisation would lead to the unjust provision of a medical genetic service.

REFERENCES

Bundesärztekammer, 1992. ' Memorandum, Genetisches Screening', *Deutsches Ärzteblatt* 89, 25126, pp. 1433-1437.

Bundesminister für Forschung und Technologie (ed.), 1990. *Die Erforschung des menschlichen Genoms. Ethische und soziale Aspekte,* Campus Verlag, Frankfurt and New York, pp. 198-203.

Berufsverband 'Medizinische Genetik' 1990. 'Stellungnahme zu einem möglichen Heterozygoten - Screening bei Cystischer Fibrose', *Medizinische Genetik.,* 2/3, pp. 6-7.

Coutelle, C., 1994. 'Proceed with caution - but proceed !', *Human Genetics,* 94, pp. 28-30.

Gille, Ch. et al., 1991. 'A pooling strategy for heterozygote screening of the 508 cystic fibrosis mutation', *Human Genetics,* 86, pp. 289-291.

Kruip, St., 1993. 'Stellungnahme zum Mucoviscidose-(CF)-Heterozygotentest', *Medizinische Genetik,* 4, pp. 406-407.

Pander et al., 1992. 'Heterozygoten-Testung bei Mucoviscidose. Eugenik, Prävention oder Instrument der humangenetischen Beratung?', *Deutsches Arzteblatt,* 89, pp. 2786-2790.

Schmidtke, J., 1994. 'Proceed with much more caution', *Human Genetics,* pp. 25-27.

Schmidtke, J., 1996. 'Vererbung und Vererbtes', *rororo Sachbuch,* 1690, pp. 231-251.

Schmidtke, J., 1995. 'Die Indikationen für die Pränataldiagnostik müssen neu begründet werden', *Medizinische Genetik,* 1, pp. 49-52.

Schmidtke, J., and Vogel, W., 1990. 'Zystische Fibrose. Überlegungen zu einem Überträger-Screening', *Deutsches Arzteblatt,* 87, pp. 1867-1868.

Schmidtke, J., and Wolff G. , 1991. 'Die 'Altersindikation' ihre Abschaffung und die Folgen', *Medizinische Genetik,* 1, pp. 16-19. Discussion of this paper in *Medizinische Genetik,* 1991, 4, pp. 27-33.

Schroeder-Kurth T., 1989. 'Versorgung der Bevölkerung der Bundesrepublik mit humangenetischen Leistungen. Beratung und Diagnostik' in T. Schroeder-Kurth (ed), *Medizinische Genetik in der Bundesrepublik Deutschland.* J.Schweitzer Verlag, Frankfurt, pp. 19-47.

Schroeder-Kurth, T., 1996. 'Selbstbestimmung und Manipulation. Zu einem menschenwürdigen Umgang mit den Möglichkeiten der Medizinischen Genetik', *Berliner Medizinethische Schriften,* 11, Humanitas Verlag.

Schroeder-Kurth, T. et al., 1996. *Humangenetik und Gesellschaft,* Institut für Humangenetik, Universität Heidelberg, pp. 10-12 , 60-73.

Chapter 8

Population screening in Greece for prevention of genetic diseases

CATERINA METAXOTOU AND ARIADNI MAVROU
Genetics Unit
Athens University
Greece

1.   INTRODUCTION

Population screening for prevention of genetic diseases in Greece is focused mainly on the identification of carriers of haemoglobinopathies, since thalassemia is the major public health problem of the country.

Massive neo-natal screening for the detection of other genetic disorders where early diagnosis and treatment can prevent the establishment of incipient disease in an individual, is performed for hypothyroidism, glucose-6-phosphate dehydrogenase deficiency (G6PD), phenylketonuria (PKU) and hyper-alaninemia.

Pre-natal testing is well accepted in Greece and approximately 6.5% of Greek women undergo pre-natal diagnosis for chromosomal abnormalities (Metaxotou et al, 1997). Ultrasound screening during the 1st or 2nd trimester of pregnancy for the early detection of congenital anomalies and identification of high risk pregnancies for chromosomal abnormalities is available to all pregnant women. Lately an effort is underway for carrier detection of other monogenic disorders, such as cystic fibrosis.

R. Chadwick et al. (eds.), The Ethics of Genetic Screening, 89–93.

2.   HAEMOGLOBINOPATHIES

There are twenty thalassemia prevention centres distributed around the country which offer only carrier detection. In Greece the frequency of ß-thalassemia is unevenly distributed. There are areas in the north with low prevalence (<5%), where the expected frequency of a homozygote based on the gene frequency is 1:1660 births. On the other hand, there are certain areas in central Greece and the islands with prevalence of the thalassemia trait >15% and frequency of homozygotes as high as 1:100 births. The overall frequency of carriers in the population is 10% for all haemoglobin abnormalities and 8.5% for ß-thalassemia (Kattamis, 1983). Priority for screening is given to individuals or couples who are at immediate risk, such as young couples or pregnant women. If one partner or the expecting mother is a heterozygote then the other partner is tested. If he is also shown to be a heterozygote, the couple is referred to one of the special genetic centres where DNA analysis and precise identification of the mutation is performed. Pre-natal diagnosis is recommended and genetic counselling plays an important role, as the couples must clearly understand the risk for an abnormal child and the options available to them.

It is estimated that approximately 70% of the population has been tested and based on the gene frequency, the number of abnormal children expected to be born yearly, without prevention, is 160. As a result of screening and pre-natal diagnosis, only 10-15 affected children are born.

It must be pointed out that the screening program for haemoglobinopathies is not enforced by law but works on a volunteer basis. The effectiveness of the programme depends mainly on the public awareness of the benefits of screening. Public awareness is enhanced with various methods such as special campaigns through the media, etc.

3.   CYSTIC FIBROSIS

Cystic fibrosis is the second most common genetic disease in Greece, with a gene frequency around 5%. Unlike northern Europe the pattern of mutations is highly heterogeneous. Fifty-five mutations have so far been genotyped in 350 children with cystic fibrosis and are responsible for 90% of the cystic fibrosis transport regulator (CFTR) genes in the Greek population (Tzetis et al., 1997). Cystic fibrosis screening in the wider population has been carried out by 'cascading' through the relatives of index cases. One hundred individuals have been tested by cascade screening for the W carrier status and in the case of couples, the partners of individuals identified as cystic fibrosis carriers are also screened. Pre-natal diagnosis is offered to couples when both are cystic fibrosis carriers. This program is organised by the Unit of Molecular Medicine at Choremeio Research Laboratory in Athens. A plan to institute screening for carriers in the general Greek population, similar to the one applied for ß-thalassemia, is underway. Direct cystic fibrosis mutation analysis will offer the possibility of population-based carrier testing to healthy couples who have no family history of this disease.

4.   NEWBORN SCREENING

Mass newborn screening in Greece is done by the Institute of Child Health at Aghia Sophia Children's hospital in Athens. There are programmes for the early detection of congenital hypothyroidism, phenylketonuria (PKU), hyper-alaninemia and glucose-6-phosphate dehydrogenase deficiency (G6PD). Between 95,000 and 96,000 individuals are tested per year, on average resulting in the detection of 30 cases for congenital hypothyroidism, 6 cases of PKU and 7 of hyper-alaninemia and 1800 cases of glucose-6-phosphate dehydrogenase deficiency (G6PD) deficiency. Counselling, treatment, dietary advice etc. is offered by health professionals when a positive test result is found.

5.   PRE-NATAL DIAGNOSIS

The diagnostic procedures currently available in Greece for pre-natal diagnosis of congenital malformations and chromosomal abnormalities are fetal karyotyping, screening for biochemical serum markers in the first and second trimester of pregnancy, ultrasound examination and molecular diagnosis of several monogenic disorders (Metaxotou et al., 1997).

More than 90% of pre-natal diagnoses conducted each year in Greece are performed to detect chromosomal abnormalities, either in the 1st or the 2nd trimester of pregnancy. In 1995, 6500 women had amniocentesis for chromosomal anomalies (6.5% of all pregnancies). The main reason for referral was advanced maternal age (78%), abnormal biochemical markers

(12%) and previous abnormal child or ultrasound findings (9%). During the last two years, however, requests due to ultrasound findings continuously increased.

The lack of registries means that it is difficult to estimate the impact of pre-natal diagnosis on the incidence of chromosomal abnormalities. It can be stated, however, that among 1750 amniotic fluid samples tested in the laboratory of the 1st Department of Paediatrics in 1995, 27 abnormal fetuses were detected.

The existing legislation in Greece allows the termination of pregnancy up to the 24th week of gestation, in cases of severe, non treatable genetic disorders. The decision to terminate an abnormal pregnancy is always taken by the family and is supported by the doctor and genetic counsellor.

Greek women seem to be aware of the possibility of pre-natal diagnosis. In order to explore the attitude and use of pre-natal diagnosis services, a national survey was performed in 1995 for the 1st Department of Paediatrics among 3000 Greek women (Mavzou et al., 1997). As the study showed, 52% of the respondents, regardless of age, were adequately informed about the possibility of pre-natal testing about genetic disorders, while 48% had either superficial knowledge or no knowledge at all. The majority said that they were informed by more than one source, mainly their doctors and the media. There was a positive correlation between awareness and acceptance of pre-natal diagnosis and the social, educational and financial profile of the women.

Almost all pregnancies are tested for early detection of congenital abnormalities or identification of high risk pregnancies for chromosomal abnormalities by ultrasound during the 1st and 2nd trimester (Metaxotou et al., 1997). There are serious problems, however, regarding quality control and interpretation of ultrasound findings especially in some private Centres around the country. Our future efforts should therefore concentrate on establishing professional guidelines which must be followed by all centres involved in pre and postnatal screening for genetic disorders.

REFERENCES

Kattamis, G., 1983. 'Massive screening for prevention of thalassemia and other genetic diseases', *Pediatriki Suppl.*, 46, pp. 279-284.
Mavrou, A., Metaxotou, C., Trichopoulos. D., 1997. 'Awareness and Use of Pre-natal Diagnosis among Greek Women. A National Survey', *Prenatal Diagnosis.*, (forthcoming).

Metaxotou, C., Mavrou, A., Antsaklis, A., 1997. 'Pre-natal diagnosis services in Greece', *European Journal of Human Genetics,* (suppi l), pp. 39-41.

Tzetis, M., Kanavakis, E., Antoniadi, Th., Doudounakis, S., Adam, G., Kattamis, C. 1997. 'Characterization of more than 85% of cystic fibrosis alleles in the Greek population, including five novel mutations', *Human Genetics,* 99, pp. 121-125.

Chapter 9

Ethics and genetic screening in the Republic of Ireland

DOLORES DOOLEY
Department of Philosophy
National University of Cork
Ireland

Genetic screening in the Republic of Ireland is in the early stages of organisation as a service under the Department of Health. The prioritising of genetic screening as a nationally available service, looked on with some enthusiasm a decade ago, seems to be halting. At least two dominant factors may partly explain this pulling in of reins: Department of Health financial stringencies *and* ethical debates concerning genetic screening. A brief history is necessary to focus the relevance of these two factors in the development of genetic screening.

Some background about the size and population of Ireland is in order to clarify the structures of Health Care Services. The Republic of Ireland consists of twenty six counties out of thirty-two comprising the whole island. Six counties of the North form Northern Ireland and comprise part of the United Kingdom, and hence subject to different policy and legislation on genetic screening. The population of the Republic is approximately three million. Dublin is the capital city of the Republic and includes well over one half of the population of the country. The health care system in Ireland is a two tier system of public health care (complete eligibility for approximately 35% of the population) and limited eligibility for the remainder of the population. Private health care insurance is an option chosen by a high percentage the Irish population.

As is the case in many other countries, medical genetics services in Ireland have been established largely in an ad hoc manner with one notable exception. For over forty years, a national screening service for newborn babies has been in operation at the Children's' Hospital, Temple Street, Dublin. This centre

R. Chadwick et al. (eds.), The Ethics of Genetic Screening, 95–104.
© 1999 *Kluwer Academic Publishers. Printed in the Netherlands.*

provides diagnosis and treatment of five inherited metabolic disorders of which phenylketonuria (PKU) is the most common and best known. The principle followed in the genetics service at Temple Street is that only those diseases which are treatable under present scientific knowledge, are tested for in newborns on a national basis. Parents of newborns are encouraged to participate in this testing since the treatment benefits are available. There are no reported cases of non-consent from parents.

In recent years, there have been some individuals doing genetic testing and counselling on a private basis, but interruption in these disparate services occurred when interested and trained individuals resigned or retired their posts for a variety of reasons. In addition, private services do not address the needs of the public sector, an inequity which does not sit comfortably with the Department of Health, lay persons or medical practitioners. Until recently, this ad hoc and uncoordinated situation in genetics services did little to further national coordination, continuity of service or development of national guidelines for genetics services.

The need for a coordinated and regulated medical genetics services is acknowledged as a priority among medical consultants and researchers in the Republic of Ireland. For example, research to date indicates that cystic fibrosis is the most commonly known inherited genetic disorder in Ireland. Approximately 1 in 16 carries the mutated gene and about 1 in 17,000 live births are affected by it. Since genetic therapy is seen as a real but distant possibility, researchers at University College, Cork are working on a way of bypassing the genetic defect for cystic fibrosis (CF). The goal is to open up other pathways for salt movement in the cells and to identify the molecular mechanisms in cells for controlling salt transport. Extensive genetics research is also underway at Trinity College, Dublin, University College, Dublin, University College Cork, University College, Galway. Among the diseases being studied closely are malignant hyperthermia, cystic fibrosis, colo-rectal cancer and central core disease.

1.1   Introducing a National Genetics Centre

With dissatisfaction among clinicians and researchers at the uncoordinated and minimal clinical genetic testing and counselling, a systematic review of the requirements of a genetics service was carried out by the Department of Health, between 1988 and 1990. Professionals, who were largely drawn from pathology and genetics, comprised a study committee to examine the needs for such a service, terms of reference for such a clinical genetics service and presently existing facilities. Submissions were invited and received from most health boards and hospitals throughout the country. The Committee had broad parameters:

1. to recommend how a medical genetics service should be organised to provide an adequate clinical service;
2. to identify the laboratory support required;
3. to recommend appropriate links between the clinical service and teaching institutions.

The process of two year study was disciplinarily narrow in its consultative process. The committee was weighted heavily towards medical representation and consisted of the Deputy Chief Medical Officer from the Department of Health, six consultant specialists in pathology, paediatrics, and genetics and three health care administrators from the Department of Health. Extensive information-gathering followed with visits to genetics research centres. Written submissions were invited from units including hospitals and universities, health boards, the Royal College of Psychiatrists, the Institute of Obstetricians and Gynaecologists and the National Association for the Mentally Handicapped. This consultation process, coupled with the collective knowledge and experience of the committee members, constituted the basis on which the Committee's report was based.[1] "The proposed medical genetics service should be provided in the context of a national policy and in a co-ordinated manner to serve the needs of the people of the country as a whole" (RGS, 1990 p. 34). The report recommended that a co-ordinating committee be established to ensure close co-operation and collaboration in the provision of a "comprehensive medical genetics service of the highest quality to the whole country" (RGS, 1990, p. 34). The membership of the co-ordinating committee would consist of: the consultant medical geneticists; the laboratory heads and representatives of the management of each medical genetics service provider.

Multiple functions specified in the job description for the clinical geneticist made some observers and the medical geneticist sceptical about the likelihood of successful execution of all the assigned tasks. These included: teaching postgraduates; undergraduates and paramedics; engage in research; compile a Regional Genetic Register; counsel patients and/or their parents and relatives and to liase with those responsible for cytogenetic, biochemical and molecular genetics and other laboratory genetic services. The job description expanded beyond even these terms and would prove a Herculean task for a single clinical geneticist to carry out effectively. The concept of a National Centre with a fundamental role for co-ordinating work throughout the country with a population of 3,000,000 may have had serious structural design faults and inadequate financial commitments from the outset.

The Department of Health review group concluded that more systematic organisation and provision of medical/clinical genetics services was clearly necessary. The questions that had to be decided were timing of developments, finances, personnel required, number of centres and location for genetics centres. As a result of this review, the Department of Health decided, in 1990 under the direction of the then Minister for Health to establish the first national and public genetic testing and counselling service. In July of 1994 this service was established in Dublin. One medical geneticist was appointed. The development was publicised in a low-key manner, revealing a concern that some interest groups might object to all screening services out of the fear that routine ante-natal testing would become widely available. Among many practising doctors, the Unit promised a public service that had never existed *in any co-ordinated way* in the Republic with a population of approximately 3,000,000. By January of 1995, eleven scientists had been hired and the amount of samples coming to the Centre for genetic analysis and/counselling approximated 2,000 by the end of 1995 with anticipation of increasing numbers each year. The national interest in such a service is evident if one were to judge from this initial use of services. In planning the National Unit, it was acknowledged that, for some years, a minimum scale commercial service in genetic testing had been available on a fee per sample basis in University College Galway and Trinity College, Dublin.

1.2   Omitting inter-disciplinary in-put

In its organisation, the model for structuring the Genetics Unit and co-ordinating the country's services was based on a predominantly scientific model, the limits of which are evident in the following points (RGS, 1990). Neither the study committee which came to recommend the Genetics Unit nor the Committee for on-going Co-ordination was given any authority, nor it seems sought any brief to involve members outside scientific disciplines nor to develop underlying normative philosophies or policies which would help to clarify the ethos, purpose and limits of the Genetics Unit. If such inter-disciplinary dynamic had existed from the formative stages, public debate would be at a much more advanced stage than it is at present. It is not too late to broaden the co-ordinating committee to take in a much wider diversity of viewpoints, queries, suggestions about genetic service developments.

Perhaps with this in mind and to the credit of the planning committee, a 'Users' Forum' was recommended. This was meant to comprise representatives of patients who might come from the Inherited Disorders Organisation, hospital administrators, medical schools, research institutions, consultants, general practitioners and other 'users' of the medical genetics services. But an opportunity was left unexploited to involve a wider representation of lay users. This users committee has not materialised but its description suggests again a focus on genetics users as medical personnel or administrators working within a medical service. What became apparent was the absence in all of these structures of lay persons' voices and input from disciplines such as social science, ethics and theology. The rather exclusive 'scientific' model determining membership and organisation indicates that neither policy formation nor statement of a normative philosophy for the service was seen as essential. The scientific 'expertise' focus left seriously under-represented the lay persons who were the presumed beneficiaries of this genetics service.

It is clear and not surprising that normative assumptions guided the final formulation of the report. Technological and scientific advances including genetics services are always used in a particular social context and cannot be neutrally understood without awareness of a context. Information is rarely neutral and organisational decisions are no more so. In examining the evolution of genetic services in Ireland, it is increasingly evident that information is prepared and presented and often with-held or circumscribed by institutions or individuals with their own interests and values.[2] An example is the choice of the working committee which drafted the plan for the Genetics Service to include the following specific points, signalling an adoption of positions about normative and legal issues they deemed essential in the process of development of a National Genetics Centre. The presence of normative parameters is not the problem but rather the complaint is with their formulation within a profoundly limited representation of disciplines and genetics users drawn from a range of social contexts. Explicitly stated::

> Genetic counselling assists individuals to make important decisions particularly about reproductive options. Important moral, ethical and legal issues are raised in the practice of medical genetics. It is an accepted principle that genetic counselling should be non-directive. (RGS, 1990, p.8).

This statement reflects a genuine concern that the genetics service not become a centre for abortion advice and referral which might follow on ante-natal testing. This normative provision is a statement which reflects the aim of protecting a strong pro-natal philosophy. Such a philosophy has been traditionally voiced as an essential part of Ireland's dominant religious ethos. The committee members clearly wanted to reinforce this ethos by claiming that:

> Of course any decisions by those concerned must reflect the prevailing ethical, legal and constitutional position. (RGS, 1990, p.8).

The planning committee assumed a prevailing ethical, legal and constitutional position yet, even in 1990 when the report was drafted, many 'ethical, legal and constitutional positions' assumed to 'prevail' were recognised as culturally contestable. The emerging value pluralism within the culture of the Republic is being further adjudicated with the emergence of issues revolving around the absence of adequate and accessible genetic services.

1.3   Essential need for genetic counselling

The Department of Health report emphasises the necessity for the genetics service to be equipped with counsellors for many sensitive testing situations requiring pre- and post-test counselling (RGS, 1990). Discussions with many involved in genetics research and testing made it very clear that there is strong concern about the continuing lack of an infrastructure of counselling and support systems which would be essential for any testing or screening processes.   Counselling services emphasise: the provision of information, pre- and post-test counselling, analysis and communication about test results especially with reference to explanation of significance of the test results in terms of prognosis, available treatments, and options available to patients. The genetic counselling is non-directive in its underlying philosophy. Where counselling precedes or follows ante-natal testing, the medical genetics counsellor needs to clarify the fact, mentioned above, that presently Ireland does not have legal provisions for abortion should a woman wish to choose this option.

A lacuna in provision of trained counsellors has been identified by professionals across disciplines as a serious deficit in any adequate infrastructure for a National genetics centre. There are currently no training programmes for genetics counsellors available in Ireland. The admitted lack of counselling skills on the part of general practitioners and consultant specialists makes them reluctant to engage in pre-or post-test counselling either with patients or relatives of patients. The net result of this lack of counselling provisions, coupled with a recognition of its necessity, is that genetics services are not publicly visible and often are not being recommended by general practitioners or medical specialists. The infrastructure of trained counsellors is a primary need in the future developments and cannot wait for expansion of genetic testing and screening but needs to be developing in tandem - again as part of the process. In addition to the national genetics centre in Dublin now in operation, other sites for genetics centres are envisaged over the long term, possibly situated in the cities of Galway and Cork. If such expansion occurs, it will only aggravate the problem of non-availability of trained counsellors and the apparent lack of urgency in the universities to introduce such training programmes. Guidelines on genetics testing and counselling which are unique to Ireland do not presently exist. Nettles of ethical controversy would need to be grasped and debates encouraged if such policy guidelines are to develop at a National level. Ironically, it took the cloning of the famous or infamous sheep 'Dolly' for the Department of Health to publicly announce that it will be constituting a national ethics committee to review newly emerging ethical issues in medical technology and treatment. Because of the operational challenges mentioned above related to availability of trained genetics counsellors, the need to articulate guidelines and policies on genetic testing and informed consent provisions, there is guarded optimism about progress in public use of the National co-ordinating unit. It is guarded because those involved in any institutional decision-making dependent on Government Departments know that decisions about allocation of health resources are complicated both by politics and economic priorities and the latter are themselves so often politically and ethically determined priorities.

1.4   Medical Council guidelines related to genetics

The Medical Council of Ireland, the registration council for doctors in the Republic, issued a revised edition of its *Guide to Ethical Conduct and Behaviour* in early 1994.[3] There it endorses guidelines for In Vitro Fertilisation which were formulated by the Irish Institute of Obstetricians and Gynaecologists. Among the requirements of the guidelines, it is stipulated that only ova and sperm from the counselled couple will be used and all fertilised ova resulting from the process must be implanted. The implications of the guidelines are that, under 1994 guidelines, freezing of fertilised ova is not permitted and gene research which results in destruction of early formed embryos is disallowed. Consistent with this, genetic testing on embryos with a view to deciding whether to implant or not is not sanctioned. These same Medical Council Guidelines (1994) also speak to the question of required consent by patient to the release of any genetic information culled on that patient. The Council directives make explicit that "any request by a third party for information about a patient should be refused unless consent by the patient has been given."[4] It follows that neither insurance companies nor businesses can, without their client's consent, receive genetic information held by any medical practitioner or clinical genetics centre regarding that individual. In this context, it should also be noted that businesses and insurance companies are not legally entitled to require genetic testing of their employees or clients as a basis for deciding on employment or insurance agreements.

1.5   Debating 'Normality and Abnormality'

At this point in time it would be an exaggeration to say that there is a robust national 'debate' on the delineation between normality and abnormality especially as those categories might play a normative role in discussions about advisability or otherwise of genetic screening. Where debate about 'normality' and 'abnormality' exists, it has been largely in the institutional contexts of scientific research, government (Health Department) or medical ethics courses in universities and nursing schools. However, an awareness of genetics and its potential is increasing with researched television discussions being a popular format for broadening interest and questioning in the Republic of Ireland.

While Ireland is gradually becoming a more ethically diverse country, this diversity is always against the backdrop of a strong heritage of religious, moral and cultural values. The heritage is such that it continues to question the value of defining some conditions as 'normal' and others 'abnormal' especially and most importantly where those definitions may have implications for protection or nurturing of human lives. There is likewise a long-standing awareness of the dangers of social and individual 'stigmatisation' resulting from many diagnostic categories especially within psychiatric, disability or rehabilitation units. A recently published Report of the Commission on the Status of People with Disabilities reinforces a growing politicisation and education of a wide diversity of citizens who are broadly described as 'disabled'. The Commission's first statement about the legal status of disabled is drawn from reflects the points made above:

> Definitions of disability should use language which reflect the right of people with disabilities to be treated as full citizens: all definitions of disability should be reviewed and inappropriate and offensive language replaced.[5]

As a result of such apprehensions about discrimination, efforts to specify 'normal' and 'abnormal' on the basis of one's genetic history, one's diseased condition or one's carrier status tend to meet with considerable cautious scepticism. In brief, *if* there is no gene therapy, counselling or psychological assistance to help individuals understand the significance and consequences of these life markers (genetically normal or abnormal), then there remains a strong presumption in the Republic that the markers should not be applied and services which would impose these markers not be encouraged. Against this cultural background, the process of determining the boundaries of 'normal' and 'abnormal' and problems in those determinations are likely to come under increased public discussion and challenge as awareness of expansion in available genetics services continues.

1.6   Social response to developments?

There has been little social response to these developments because at present there are only subdued beginnings of national publicity given to genetic screening potential and its implications. The national service in Our Lady's Children's Hospital (Crumlin Street Hospital) in Dublin will elicit more discussions of the meaning and importance of non-directive counselling, the facts and treatment options available for various inherited genetic disorders and the myriad of social, familial and psychological implications for individuals diagnosed as having genetic diseases or having carrier status.

Perhaps most audible in the public sphere of debate is the close link perceived between Ireland's laws on abortion and the increasing requests for ante-natal testing services. Ireland does not have legislation which allows for legal abortion. This legal vacuum surrounding abortion is in spite of a Supreme Court decision *Attorney General vs. X and Others* in 1992 which argued that abortion is legal in extremely limited cases. However, subsequent governments have not taken up the challenge of trying to formulate acceptable legislation. Because of the unavailability of abortion in the Republic of Ireland, it has continued to be argued that availability of ante-natal testing specifically amniocentesis is an incoherence in the system or a serious moral anomaly if the choice of abortion is not available within the jurisdiction of the Republic. Politico-ethical and legal positions within the Irish culture have not, to date, actively mobilised for provisions of ante-natal testing which would include the option of deciding on the continuation of a pregnancy. This deficiency in the health care services has, for some time, been recognised, at professional level of obstetrics and gynaecology and in women's groups, as needing change. However, the Genetics Centre described above is very quiet about the prospect of providing amniocentesis - again for political and ethical reasons which seem to help maintain an unpoliticised management of the Genetics Unit. In the absence of ante-natal testing services, Irish doctors have mainly utilised services available in Belfast and London.

Since the early 1980's, recurring national debates about abortion information and possible legislation have proved divisive and painfully educational at the same time. Currently there is a clearer positive response to the constructive potential possibilities of genetic testing aimed at helping couples in their decision-making about beginning a family.[6] There is more of a mixed reception for making available ante-natal testing of pregnant women with a view to providing referrals for abortions if the women so choose. On this issue and, with the background culture and history in reproductive ethics in the Republic, the debate about 'normality' and 'abnormality' brings responses of considerable caution. What indicators will be considered 'serious enough' to warrant an abortion? Who should be entitled to make these decisions about an abortion on the basis of genetic indicators? There is a rather robust awareness that a decision for limited abortion of the type recommended by *Attorney General vs. X and Others* just might constitute the first step on the proverbial moral `slippery slope'. The very recent introduction of a medical/clinical genetics service for diagnosis and counselling and growing media attention to genetics will, almost predictably and over the coming years, provoke greater social awareness and recognition of need to debate. There is a recurring public voice which argues that developments will occur anyway and the absence of visibility or lack of debate will not halt that. Many commentators in health care ethics argue that it is better to be prepared than met with an unacceptable and unprepared for *fait accompli*.

NOTES

1 *Report of the Committee To Examine Medical Genetics Service (RGS)s*, unpublished document housed in the Department of Health, August, 1990.
2 For a collection of essays which explores the many facets of genetic services and principles involved in such organisation see, Clarke, A. 1994. *Genetic Counselling: Practice and Principles*, Routledge, London.
3 The Medical Council of Ireland, 1994. *Constitution and Functions: A Guide to Ethical Conduct and Behaviour and to Fitness to Practise,* (4th edn), pp. 62-63.
4 *ibid.,* p. 32.
5 *A Strategy for Equality,* Report of the Commission on the Status of People with Disabilities, p. 15. The Commission was established on 29 November, 1993 by the then Minister of Equality and Law Reform, Mervyn Taylor.
6 There is a noticeable reluctance at Department of Health level to incorporate 'genetic services' as an explicitly mentioned provision in the programme for future services - even while a National Genetics Centre has been established. A publication of early 1990s from the Department of Health makes no mention at all of genetics services for screening or counselling. See *Shaping a Healthier Future: a Strategy for Effective Healthcare in the 1990s*, Stationery Office, Dublin. Available at Government Publications Sales Office, Molesworth St. Dublin.

Chapter 10

Genetic screening in the Netherlands
The state of the debate

ROGEER HOEDEMAEKERS
Department of Ethics, Philosophy and History of Medicine
Catholic University of Nijmegen
The Netherlands

1.   MAJOR PUBLICATIONS IN THE NETHERLANDS

The Health Council of the Netherlands has repeatedly given attention to issues related to genetic screening and testing. In 1977 a first report on this subject was published with recommendations for screening for congenital metabolic disorders (Gezondheidsraad, 1977). In 1979 and 1980 two other reports appeared which also discussed genetic screening issues (Gezondheidsraad, 1979; Gezondheidsraad, 1980). These reports dealt mainly with issues concerning individual testing and counselling. In 1988 a report on the social consequences of genetic screening and testing was published, discussing the use of genetic information by insurers and employers, patient rights and the role of public authorities (de Wert and de Wachter 1990). In 1989 The Health Council of the Netherlands published a new report, *Heredity: Society and Science* which covered the most important issues concerning the ethical, social and legal implications of genetic screening and testing (Gezondheidsraad, 1989). At the end of 1994 the Health Council issued a report presenting its stance on issues related to large-scale genetic screening (Gezondheidsraad, 1994).

Since the 1989 Health Council report the debate has intensified. In 1990 a first comprehensive publication on ethical aspects of genetic carrier screening was published (Schellekens, H. 1993). Three major political parties, the Christian Democrats (VSOP, 1992), the Social Democrats (Reinders, 1996)

R. Chadwick et al. (eds.), The Ethics of Genetic Screening, 105–118.

and the Liberals (Dees et al., 1994) have also prepared reports on this issue
and have presented recommendations. In 1993 a preliminary report focusing
on the role and function of the Law with regard to issues linked with genetic
screening, testing and counselling was published (de Ruiter and Sutorius). In
1995 a three-day public debate was organised by The Rathenau Institute on
the issue of predictive genetic screening (Rathenau Institute, 1995). Finally the
VSOP, a co-ordinating organisation of patient and parent groups has devoted a
great deal of attention to issues related to hereditary diseases and has also
contributed to the debate with a report (de Ruite and Sutorius, 1993). In 1996
a preliminary report appeared which discussed the question in how far a
restrictive policy for pre-natal screening and/or testing is morally acceptable
in order to reduce negative societal consequences of genetic diagnostic
technologies for handicapped people (Reinders, 1996).

2.   SCREENING AND TESTING PROGRAMMES

Since 1979 there have been eight Centres for Clinical Genetics operating
in the Netherlands. They have a regional function and the location of these
Centres around the country was regulated by the Ministry of Health. These
Centres carry out genetic testing and provide genetic counselling and follow-
up support at the request of referred individual clients or patients. The
indications for the various tests have been stipulated precisely and the Centres
are asked to submit annual reports on the numbers of patients and carriers
tested in the different indicated groups. There is quality control at various
levels (Gezondheidsraad, 1989). Four large-scale screening programmes are in
operation: pregnant women are tested for diabetes and ABO rhesus
incompatibility, newborns for phenylketonuria and congenital hypothyroidism
(Gezondheidsraad, 1994).

3.   THE RECENT DEBATE ON GENETIC
         SCREENING

The 1989 report of the Health Council (Gezondheidsraad 1989), was
mainly concerned with ethical and legal aspects of genetic testing on request,
concentrating on the screenee's right to information, his/her right not to be
informed, the right to confidentiality, protection of privacy and provision of

information to family-members. Among the recommendations proposed were the following:

1. There should be access to genetic testing and counselling for all;
2. Privacy should be protected and personal freedom of choice should not be violated;
3. Further research is recommended with regard to the conditions for which pre-natal diagnosis is indicated;
4. The right to be informed and the right not to know are emphasised;
5. The consent requirement for storage of genetic data into a registry was considered of particular importance. An individual must possess the right to have data removed or rendered anonymous. The person screened must be asked to authorise how his/her genetic data will be used. Cell banks should accept a code of conduct which gives heed to the donor's rights, but which also allows biomaterial to be used for the benefit of others or for scientific research.

The 1989 Health Council Report did not recommend neo-natal screening for untreatable, late onset disorders. This also applied to screening for disorders manifesting in childhood for which effective treatment was unavailable or diagnosis was not reliable. The report noted that large-scale carrier screening programmes needed careful consideration, as there may be important limitations (such as genetic heterogeneity and lack of reliable methods of detection). Large-scale programmes should only be introduced if the benefits clearly outweigh the disadvantages.

The Report also recommended that testing of people who apply for any type of insurance should be prohibited. At the same time insurance companies should be protected from adverse self-selection. Further investigation of this issue was thought to be necessary. Use of tests of genetic disposition in selection of employees was rejected. The Health Council warned against discrimination.

Wert and Wachter (1993) examined various ethical aspects of genetic carrier screening. They rejected mandatory testing, limitation of the right to reproduction and testing of incompetent children and adults in the interest of third parties only. They supported large-scale genetic education of the general public, which should enable individuals to make informed choices concerning reproduction. Much emphasis was given to the principle of respect for autonomy and self-determination. They recommended that population-based genetic screening must be acceptable to the target group and that pilot studies be performed for each screening programme to investigate possible negative consequences. Genetic screening programmes should include pre-test education, counselling, psychosocial support and evaluation. Carrier screening of children was rejected, carrier screening before pregnancy was preferred to screening during pregnancy. Neo-natal screening for sickle-cell anaemia was also recommended. Wert and Wachter considered a right not to know to be important but further consideration of this issue was recommended. Confidentiality should be violated only in exceptional cases. The individual should have optimal control about use of genetic data or biomaterial. The authors also expressed concern that prevention programmes may affect care for handicapped persons.

Three political parties have considered issues linked with genetic screening and testing. These have been compared and the most important issues can be found below in tables 1, 2, 3, 4.

Table 10.1 Criteria proposed by three Dutch political parties for genetic screening programmes

	Christian Democrats	Social democrats	Liberals
Benefits aimed at	Individual health benefits: prevention or early treatment; improved prognosis.	Reduction of frequency of genetic disorder in population. Screening must benefit individual and offspring.	Well-informed decision-making about reproduction.
Values/ principles emphasised in report.	Autonomous decision-making. Integrity of human person. Respect for and protection of human life.	Prevention of human suffering. Respect for individual norms/values. Freedom to decide in matters of reproduction.	Individual autonomy. Integrity of human mind and body. Uniqueness of individual. Tolerance and freedom of conscience.
Perceived risks/harms individual.	Social pressure. Psychological burden. Stigmatisation. Less employability and insurability. Medicalization.	Discrimination of population groups. Social pressure. Autonomous decision-making threatened, if no adequate care for handicapped people.	Invasion of privacy. Psychological burden. Social pressure caused by financial consequences for parents with handicapped children.

Table 10.1 (cont.)

	Christian Democrats	Social democrats	Liberals
Perceived risks/harms society.	Government permit may reinforce eugenic tendencies. Changing attitudes towards handicapped people.	Not discussed in report.	Discrimination. Eugenic tendencies Changing attitudes towards handicapped people.
Carrier screening	Rejected, if the only method of prevention is abortion.	No abortion for trivial genetic reasons. Doubts about large-scale genetic programs with abortion as only method of prevention.	Depends on attitude of parents towards abortion.
Pre-natal diagnosis	Rejected, if abortion is only method of prevention.		
Role of authorities	Prevention of harm to individual and society. Legislation: permit. Prevention of social pressure. government permit	Stimulation of prevention of serious genetic disease. National ethical committee (Danish model) recommended. European regulation preferred.	Stimulation of public debate/ critical evaluation of new developments. Evaluation and surveillance. Prevention social pressure

Table 10.2 Criteria/conditions proposed by Dutch political parties for implementation, counselling procedures and storage of information and biomaterial

	Christian Democrats	Social Democrats	Liberals
Access	Voluntary participation.	Voluntary participation.	Voluntary participation.
Pre-screening education	Extensive and adequate information.	Considered important.	Adequate information.
Counselling facilities		Considered imperative.	
Informed consent	Required.		Required.
Withholding results	Acceptable, if well-being of screenee is seriously harmed.		
Right to know	Yes, unless serious harm is done to client's well-being.		
Right not to know	Yes, unless well-being of client or third party is seriously harmed.	Yes, but this right cannot always be guaranteed.	Not absolute; duty to prevent suffering may outweigh this right.
Disclosure of unexpected findings		No unasked for information should become available. If this is not possible, procedural guidelines are needed before introduction of screening program.	Prevailing view accepted that agreements about what information is to be given before screening starts must be made.
Providing information to third parties	Confidentiality important, but not absolute: serious harm to third parties may outweigh confidentiality.		Confidentiality not absolute: serious harm to third parties may outweigh confidentiality.

Table 10.2 (cont.)

	Christian Democrats	Social Democrats	Liberals
Storage/control of data			Restricted to data required for aim of screening program. Client should have optimal control. Biological material remains the property of client.

Table 10.3. Dutch political parties statements on genetic screening linked with employment

	CDA: Christian Democrats	PvdA: Social democrats	VVD: Liberals
Perceived individual risks/harm	Tests not yet reliable. Psychological burden. Selection based on genetic constitution. Decreased employability. Consequences for relatives. Violation of right not to know.		Asking for genetic information is considered infringement on privacy. Decreased employability.
Perceived social risks/harm	Large scale discrimination when applying for job. Discrimination of ethnic groups, when applying for job.		Large scale discrimination against the genetically weak.
Workplace monitoring	Yes, but with conditions: - reliable, - voluntary, - aiming at detection of health-hazards, - leading to improvement of workplace conditions.		
Providing existing information when applying for job		Rejected.	Rejected.
Genetic screening when applying for job	Although justifiable, it is rejected in case of damage to future health or in case of risks to third parties. Also rejected because of serious social implications.	Rejected.	Rejected.

Table 10.4. Dutch political parties statements on genetic screening and insurance

	CDA: Christian Democrats	PvdA: Social Democrats	VVD: Liberals
Perceived individual risks/harms	Violation of right not to know. Psychological burden.		Higher premiums. Infringement of privacy
Perceived social implications	Discrimination. Increased reluctance to ask for genetic tests.		
Providing genetic information to collective health insurers.		Rejected.	Increased reluctance to ask for genetic tests.
Genetic screening for collective health-insurance		Rejected.	
Providing information for collective pension schemes	No right to ask information. No duty to provide information.	Rejected.	
Genetic screening for collective pension schemes	Rejected.	Rejected.	
Providing genetic information for supplementary (life, health, disability, pension) insurance	Restricted right to ask for genetic information: reasonable sums only. Restricted duty to provide genetic information (reasonable sums only).	Rejected.	Restricted duty to inform. Right to ask for genetic information: for high sums only.
Genetic test required for supplementary (life, health, disability, pension) insurance	Rejected.	Rejected.	Rejected.
Role authorities	Supporting self regulation.		

At the end of 1994 the Dutch Health Council published a report on social and legal implications of population genetic screening. Psychological, ethical, legal and social consequences were considered. The report identified four important psychological issues:

1. psychological factors that may influence a person's decision to accept an offer to be screened
2. psychological consequences of disclosure of screening results (positive: confidence and reassurance; negative: negative perception of self);
3. the impact knowledge of the test results may have on the lives of the individuals tested and their families and
4. psychological consequences for people who have decided not to participate (for example remorse or guilt).

In view of these possible negative consequences, pre-screening information and education about genetics was considered essential. Although the report noted that an invitation to undergo screening may make people anxious and worried, large-scale information about a genetic screening programme is justified as it is the only way to ensure that a test is available to all and not only to those who are already aware that it exists.

The Dutch Health Council stated that because of the disadvantages that may go with the introduction of a screening programme, possible benefits and potential harms must be carefully weighed before implementation of each genetic screening programme. Persons who offer the screening programme should have principle responsible for this evaluation. If a government permit is needed, the criteria should be examined by an independent body. For screening programmes not falling within the scope of the Population Screening Act the Council also recommended examination by an independent body.

Voluntary participation was considered essential and the report gave much attention to threats to individual autonomy and self-determination. Many forms of social pressure that may influence the decision to participate were discussed. Social stigma and discrimination were also noted and the report remarked that uncertainty about access to insurance or employment may be a reason not to introduce a genetic screening programme. Because advantages are nearly always accompanied by disadvantages, pilot studies are required to assess whether benefits clearly outweigh the disadvantages.

Social consequences do not get much attention as the main emphasis is usually on potential harm for the individual screened or his/her family. The Council recommended that there should not be excessive emphasis on the financial benefits of a screening programme, as this may threaten voluntary participation. The Council was aware that concern about access to employment and insurance may also influence decisions to participate in a genetic screening programme. The possibility of new forms of uninsurability was considered and the Council urgently requested legislation as it thought that self-regulation was not sufficient.

The Council did not specify that screening programmes should only be introduced for serious conditions. Determination of the severity of a disorder was left to the parents and this appears to be in line with the Council's shift towards enhancing autonomy as a primary target of genetic screening and testing.

4. THE PUBLIC DEBATE

In 1994 a public debate on predictive screening modelled on the Danish consensus debates was held in The Netherlands. It included several multi-disciplinary workshops on issues like screening for multi-factorial disorders, pre-natal diagnosis, genetic screening for insurance or workplace. Social, legal and ethical implications were also discussed. Representatives of parent and patient groups and of organisations of handicapped people also came together, and a three-day conference with a panel of experts, a panel of lay-persons and others interested was held in February 1995. The public debate was concluded with a final declaration which included a number of recommendations and requirements. Major concerns were the integration and support of diseased and handicapped people in society and a tendency towards greater medicalisation. A more important role of ethicists in connection with (the planning of) pre-natal and predictive screening was envisaged and a continuing discussion about the limits of intervention during pregnancy was felt necessary. Free choice was felt to be very important and various forms of (social) pressure threatening this received much attention. Non-directiveness in counselling was not always thought possible, commercialisation of genetic screening and testing not desirable and there was great concern about uninsurability. The declaration also pointed out the importance of psychosocial support, public and professional education in genetics and a remark that cost-containment in health care should not lead to social pressure for selective abortion.

In an evaluation it was noted that this debate on predictive screening had hardly been of a public nature. There was disappointingly little response in the media, and the participants could hardly be considered representative of the general population. Interest groups, such as the VSOP, the Dutch organisation of 42 national patient and parent support groups, and the Federatie Nederlandse Gehandicaptenraad (Dutch federation of handicapped people), an umbrella organisation for disabled persons, played an important role during the debate. There was a strong feeling that during the debate ethical questions had not been adequately discussed, as these questions were often countered with scientific information, the assumption apparently being that many ethical questions were the result of inadequate scientific knowledge of the state of the art in (medical) genetics. The general impression was that during this public debate little progress had been made in finding answers for the major issues.

In the January 1995 meeting of patients, parents and handicapped people great concern was expressed about possible negative consequences of genetic screening for access to insurance. Further legislation was demanded. Great fear was also expressed by handicapped people that genetic screening would reinforce negative views towards disease and handicapped life, and would lead to more stigmatisation and discrimination. This would prevent, they felt, their further integration in society. This concern was taken up by Reinders (1996), who wrote a preliminary report for the Dutch Organisation for Bioethics about the question whether the government should limit the use of pre-natal diagnostic technologies in order to prevent negative consequences for handicapped people. In this report Professor Reinders argues that a restrictive policy is not the right approach. One important argument is that the freedom of the individual to make moral choices should not be given up and a pragmatic argument is that a restrictive policy is difficult to enforce legally, because there is always the possibility of an abortion on 'social' indication.

5.   LEGISLATION

In 1993 a preliminary report was published for the Dutch Organisation of Lawyers (Nederlandse Juristen Vereniging), exploring the role that the law could have with regard to genetic screening, testing and counselling. Counselling issues such as the right not to know and disclosure of potentially harmful genetic information were discussed at length. It was concluded that no good legal solutions could be offered for the dilemmas in counselling. The authors felt inclined to rely on more pragmatic measures such as obtaining prior consent for disclosing important relevant genetic information to relatives. The authors also expressed concern that the increased diagnostic possibilities and an increased tendency towards prevention and cost-containment would result in forms of social pressure which would threaten voluntary participation. Finally the authors noted that there was a tendency to use abortion not only as a means to terminate a pregnancy that was not desired, but also as a method of selection of foetuses with undesirable conditions.

In 1996 the Population Screening Act came into force and genetic screening programmes now fall within the scope of this act. For large-scale genetic screening programmes central government approval is required before they can be implemented. A licence is refused if screening is scientifically unsound, conflicts with regulations governing medical practice or if it involves risks for the subjects that outweigh the expected benefits. Rules for population screening programmes to detect serious diseases or abnormalities which cannot be treated are very tight. This includes pre-natal screening for genetic disorders with selective abortion as the only form of prevention.

6. CONCLUSION

In The Netherlands great emphasis is given to voluntary participation and informed decision-making, and there is much concern about threats to autonomous decision-making and free choice. Protection of the private sphere is also a major concern. In order to minimise risks for individuals, safeguards are recommended. The most important are: psycho-social pre-test and post-test guidance, no discrimination when obtaining insurance or when applying for a job. Also strict rules for protection of genetic data in registers, and creation of a social climate that favours acceptance and respect for handicapped people. Organisations of disabled people are worried that in future there will be more discrimination. They are especially worried about future discrimination by insurance companies and employers, (e.g. van Wijnen, 1995). Recent legislation requiring employers to contribute more to costs of illness and disability of employees for a certain period has also enhanced their concern that there will be fewer, not more, job opportunities for individuals labelled disabled.

REFERENCES

Dees, D.J.D., List, G.A., van der List, E.G. Terpstra, 1994. *Gentechnologie, een liberale visie*, Prof. Mr. B.M. Telderstichting, Den Haag.
Gezondheidsraad, 1977. *Advies inzake Genetic Counseling*, Gezondheidsraad, Den Haag.
Gezondheidsraad, 1979. *Advies inzake Screening op Aangeboren Stofwisselings-ziekten*, Gezondheidsraad, Den Haag.
Gezondheidsraad, 1980. *Advies inzake Ethiek van de Erfelijkheidsadvisering en Genetic Counseling*, Gezondheidsraad, Den Haag.

Gezondheidsraad, 1989. *Erfelijkheid: Wetenschap en Maatschappij: Over de Mogelijkheden en Grenzen van Erfelijkheidsdiagnostiek en Gentherapie*, Gezondheidsraad, Den Haag.

Gezondheidsraad, Commissie Screening Erfelijke en Aangeboren Aandoeningen, 1994. *Genetische Screening*, Gezondheidsraad, Den Haag.

Rathenau Institute, 1995. *Predictive Genetic Research, Where are We Going? A Report to Parliament*, The Rathenau Institute, Den Haag.

Reinders, J.S., 1996. *Moeten Wij Gehandicapt Leven Voorkomen? Ethische Implicaties van Beslissingen over Kinderen met een Aangeboren of Erfelijke Aandoening*, Nederlandse Vereniging voor Bioethiek, Utrecht.

de Ruiter, J. and Sutorius, E.P.R., 1993. *Manipuleren met Leven*. Handelingen Nederlandse Juristen Vereniging 123e jaargang/1993-1, W.E.J. Tjeenk Willink, Zwolle.

Schellekens, H., 1993. *Genetisch Onderzoek en de Mens*, PvdA verkenningen 1993, P.v.d.A, Amsterdam.

van Wijnen, A., 1995. De maatschappelijke effecten van genetisch onderzoek. *Medisch Contact*, 50, pp. 1192-1195.

VSOP, 1992. *Ethiek en Erfelijkheid*, VSOP, Soest.

Wert, G.M.W.R. de en de Wachter, M.A.M., 1990. *Mag ik uw genenpaspoort? Ethische aspecten van dragerschapsonderzoek bij de voortplanting*, Ambo, Baarn.

Wetenschappelijk Instituut voor het CDA, 1992. *Genen en Grenzen, een Christen-democratische Bijdrage aan de Discussie over de Gentechnologie*, Wetenschappelijk Instituut voor het CDA, Den Haag.

Wetenschappelijke Raad voor het Regeringsbeleid, 1988. *De Maatschappelijke Gevolgen van Erfelijkheidsonderzoek*, SDU, Den Haag.

Chapter 11

Genetic screening and genetic services in Slovakia

VLADIMIR FERAK
Department of Molecular Biology
Comenius University
Bratislava, Slovakia

1.   HISTORICAL AND DEMOGRAPHIC BACKGROUND

Before attaining independence in 1993, Slovakia existed as a republic within the Czechoslovak Federation. Slovakia inherited most of its laws and other regulations governing medical practice, including those concerning genetic screening and genetic services, from the former Czechoslovakia. The same holds true for the Czech Republic; and hence the situation in both these countries is very similar.

1.1   Basic demographic data

Toward the end of 1994, Slovakia had a population of 5,390,000, according to a survey by the Statistical Institute. Administratively, the country is divided into 38 districts with Bratislava (440,000 inhabitants) as a capital.

R. Chadwick et al. (eds.), The Ethics of Genetic Screening, 119–127.

Children under 14 represented 23.5% of total population, the reproductive age population was 59.1%, and the population in post-reproductive age formed 17.4% of the total. Women of reproductive age (14-49 years) represent 50.5% of the total female population.

The natural annual population growth was 3.9 per 1,000 inhabitants (in 1993), and the birth rate was 1.39. Both these indices have tended to decrease steadily since 1975. In 1975, there were 100,000 live births falling to 73,000 in 1984.

In 1993, there were 45,000 abortions registered in Slovakia, of which 90% were artificial terminations of pregnancy. This number, divided by the number of live births gives a 60% abortion ratio, and an absolute value of 34 abortions per 1,000 females of reproductive age. Within the last 5 years, the number of abortions has decreased by 5 to 9 per cent per year. Currently there are just under 40,000 artificial terminations of pregnancy per year, out of which only about 4,000 are due to health reasons (e.g. genetic disease).

Gross mortality rate in Slovakia corresponds to 10 deaths per 1,000 inhabitants (9.9 in 1993). Children under 1 year of age contribute to the gross mortality rate by 1.5%, and the productive age population by 18.5%.

The infant mortality rate also tends to decrease steadily, with the values slightly over 10 per 1,000 (12.6 in 1992, 10.6 in 1993, and estimated 9.9 in 1994). The perinatal diseases cause, as a rule, 50 to 55% of deaths in children under 1 year of age. The second largest group of death causes in this age cohort are congenital anomalies (25%). Annually, approximately 200 children with a congenital anomaly per 10,000 live-born children are registered.

The mean life expectancy at the time of birth was (in 1994) 67.6 years for males, and 76.3 years for females.

1.2 Health establishments, health services, and medical education

In Slovakia, the 'socialist' state-controlled and state-run type of health care and health facilities (inherited from the previous communist regime) still prevail, in spite of attempts to establish a private sector within the health care system. The private sector is restricted mainly to stomatological care and pharmaceutical services.

In 1994, there were 82 hospitals in Slovakia (all of them state-run or state-subsidised) with 42,000 beds i.e. 80 beds per 10,000 inhabitants. The Ministry of Health registers. 17,000 physician's posts, i.e. about 32 physicians per 10,000 inhabitants. The total number of persons employed within the health service sector is approximately 125,000 (less then 10% of them by private organisations).

There are three Faculties of Medicine within Slovak Universities (Bratislava, Martin, Košice) with a total of 4,500 medical students. About 500 to 600 medical doctors graduate from them each year. The study of medicine usually takes six years.

1.3 Medical genetics: establishments and education

Medical genetics services are provided by the departments of medical genetics belonging to the University Hospitals and other larger hospitals. At present, there are 12 departments of medical genetics in 11 different cities (two of them in Bratislava), with a total staff of 120 (21 of whom are medically qualified doctors, and 35 biologists or biochemists with an M.Sc. degree). Most departments of medical genetics only provide genetic counselling and basic cytogenetics.

Medical genetics is not formally taught as a regular subject of the university curriculum at any of the Faculties of Medicine in Slovakia. Medical students only receive elementary teaching in human genetics within the subject of Biology during the first two study years. The absence of medical genetic education within the medical curriculum has profound negative consequences for the quality of genetic services within the health care system in Slovakia.

At the post-graduate level, medical genetics is one of the options for physicians attending postgraduate medical education courses. Medical genetics represents one of the so called second-level attestations, attainable for those who had gained their first-level attestation in either paediatrics, internal medicine or gynaecology and obstetrics. So far, not more then 25 physicians have specialised in medical genetics within the postgraduate medical education programme in Slovakia. Most of them are now active as consultants at the departments of medical genetics.

1.4 Medical genetic services

The scope and extent of medical genetic services within the health-care system in Slovakia is regulated by the 'Conception of Medical Genetics Act' that was passed in 1975 and amended in 1988. According to this regulation, the medical genetics services should be provided by a network of departments of medical genetics, which should fulfil the following tasks:

1. to perform the diagnosis, including pre-natal diagnosis, of genetic diseases by means of cytogenetic, molecular, biochemical and other relevant methods;
2. to carry out early detection of affected individuals and of carriers;
3. to accomplish genetic counselling in the affected families;
4. to take part in preventive measures, and in the treatment of genetic diseases;
5. to establish genetic family registers;
6. to conduct research in the field of medical genetics;
7. to take part in postgraduate medical education in the area of medical genetics.

In practice, the departments of medical genetics fulfil these tasks, but in most of them, the level of services provided is rather low, mainly due to inappropriate staffing and equipment, and to limited financial resources. All of them provide basic counselling and cytogenetics, but only four of them regularly perform pre-natal (cytogenetic) diagnoses, and none of them is able to perform DNA-based diagnosis. Molecular genetics is provided almost exclusively outside the network of departments of medical genetics: at the Faculty of Natural Sciences, and by a private company. A nation-wide genetic family register has not yet been established.

1.5   Pre-natal genetic diagnosis

The first pre-natal genetic diagnosis in Slovakia was carried out in 1980. Now, pre-natal diagnosis is well established at four different departments of medical genetics, though the DNA analysis is performed outside the departments (see above). DNA-based diagnosis is performed for: cystic fibrosis, Duchenne muscular dystrophy, haemophilia A and B, phenylketonuria, Huntington's disease and Fragile X syndrome. This service is accessible for families from all districts of Slovakia.

So far, only a limited number of DNA tests have been performed on children for adult diseases (two families with Huntington's disease), and the internationally recommended ethics guidelines were followed.

1.6 Neo-natal screening

At present there are only two nationwide neo-natal population screening programmes in Slovakia: phenylketonuria (PKU) and congenital hypothyroidism (CH). For both of them, the laboratory tests are carried out in a single screening centre in Banská Bystrica (Central Slovakia). Both programmes screen nearly all newborns.

The PKU screening has been running since 1973 (before that year it had run on a restricted scale), and is very efficient (as yet, no known PKU patient has been missed by the screening programme). Over two million tests has have been performed so far. The population prevalence of PKU in Slovakia is similar to that in most European countries (approx. 1:8,000), but the spectrum of mutations at the phenylalanine hydroxylase locus differs considerably. Five mutations, all of them detectable by means of a PCR-based assay, account for over 80% of all PKU mutations, which makes this disease a good target for direct DNA-based diagnosis and carrier detection in Slovakia.

The CH screening programme started in 1985 on a nation-wide basis, after having run for five years on a narrower scale (Central Slovakia only). So far, about 780,000 tests have been performed, and 122 cases of CH have been detected. The screening frequency of CH in Slovakia is 1:5,000, about twice the actual frequency of 1:12,000 - 15,000 (due to the high rate of false positive results).

No other genetic diseases are screened for in the neo-natal period, mainly because those which are screenable and treatable are, as a rule, very rare or virtually absent in Slovakia (haemoglobinopathies, thalassemics, Tay-Sachs disease etc.).

1.7 Carrier screening

Carrier screening for cystic fibrosis (CF) and other genetic diseases, for which the DNA-based carrier detection is feasible, have not yet been considered in Slovakia. This is in part due to high costs of the DNA-based screening and lack of laboratory facilities. In addition, in the case of CF, the most frequent mutation accounts for as little as 55% of all CF mutations in the Slovak population.

1.8   Pre-natal screening

All pregnant women aged 35 or more are offered pre-natal diagnosis for chromosomal anomalies and alpha-fetoprotein (AFP) determination. In most districts of Slovakia, AFP determination in pregnant women is performed regularly, though no centralised screening programme exists. These local programmes cover over 50% of the population, but no data are available so far to provide a more accurate estimate.

1.9   Pilot studies

In 1970s and 1980s, neo-natal screening for alkaptonuria was carried out in several districts of Central Slovakia by a research laboratory at the medical faculty in Martin, Central Slovakia. Alkaptonuria is a 'Slovak' disease in a sense that its incidence in Slovakia, and particularly in some isolated populations of Central Slovakia is much higher than the European average.   This screening programme determined an incidence of alkaptonuria of 1 in 20,000 within Slovakia, and helped to ascertain the affected families. However, it was eventually discontinued, since the disease is effectively incurable.

There are some pilot studies on familial hypercholesterolaemia on a very restricted local scale, but these studies are case-ascertainment rather than screening.

No other genetic screening programmes are currently being considered in Slovakia.

1.10   Genetic registers

No centralised genetic family register has been established in Slovakia. Individual departments of medical genetics keep their own local family records, but these are not linked to each other. There is, however, a centralised nationwide Register of Congenital Anomalies. This register serves as a source of basic information on children with congenital anomalies for individual departments of medical genetics, which invite the parents of these children for genetic examination and counselling.

1.11   Other genetic services

Since 1992, DNA typing for forensic and crime investigation purposes has been conducted in Slovakia by non-medical establishments. Forensic DNA-based identification is partly performed by the DNA laboratory of the Criminalistic Institute, and partly by independent private experts (court experts). They also carry out the DNA-based paternity testing for the courts (the law does not permit such testing to be performed for private purposes). Several hundreds DNA-based paternity test have been performed so far, and the demand for them is rapidly increasing.

1.12   Genetic societies and Patients Associations

The Slovak Society of Medical Genetics was established in 1969, and it now has approximately 150 regular members. It organises scientific meetings and serves as an advisory body for both the Ministry of Health and individual departments of medical genetics.

In Slovakia, there is almost no tradition for public associations of patients. This is mainly due to the fact that under the communist regime it was very difficult (though not impossible) to establish and run such a body. Nowadays several such associations are operating, particularly that for patients (and their relatives) with haemophilia, and for patients with Marfan syndrome. However, their general impact is low, they are not yet well developed, although the situation is improves rapidly.

In 1995, a foundation 'Heredity and our Health' was established, aimed at collecting financial contributions and gifts, and allocating the money to those areas of the genetic research and medical genetics practice, which most needed support.

1.13   Genetically distinct populations in Slovakia

Areas are known to exist within Slovakia with increased incidence of some genetic diseases (of which alkaptonuria is the best known example), but these regional differences are disappearing rapidly. There is one exception: a distinct population of Roms (Gypsies). The estimated number of Roms in Slovakia is 300,000, comprising over 5% of the total population. They are of Indian origin, coming from India about 800 years ago. There has been little integration with other Slovak communities, hence, their gene pool is significantly different from that of the rest of the population. Some genetic diseases are known to be much more frequent in Roms than in non-Roms (PKU, SMA, autosomal recessive congenital glaucoma, Alport syndrome), whereas others are almost absent (e.g. cystic fibrosis).

PKU in Roms represents a specific problem, as far as genetic screening is concerned. Frequency of this disease is almost 10 times higher in Slovak Roms than in the rest of the population. Since a single PKU mutation (R252W) accounts for over 90% of all PKU mutations in this particular population, and this mutation is easily detectable by a simple PCR-based test, PKU is a good candidate for a carrier screening programme in Slovak Roms. However, due to a very low average educational level in Roms, most couples from this population do not have enough educational background to cope with the information derived from such a programme. This can be demonstrated by the results of the ongoing PKU screening programme: most Rom children with PKU remain untreated in spite of the fact that they have been detected by screening, and the responsible district physicians do their best (as a rule) to convince their parents that their child would become seriously affected if they do not strictly follow the PKU diet and other measures.

Introduction of population-restricted carrier screening for PKU in Roms would probably also meet other difficulties, particularly the reluctance of Roms to be treated in a different way from other ethnic groups. Most Roms are very sensitive to what they may perceive as a sort of racial discrimination.

2.   ETHICAL ISSUES

In Slovakia, genetic counselling has a tradition of being non-directive, which may be surprising in a country which has experienced a very directive political regime for over 50 years. In spite of this, most couples opt for the termination of pregnancy if the probability of a seriously affected child is high enough (however, no statistical data are available).

The impact of genetic screening programmes, and of medical genetics services in Slovakia is undermined by the inadequate knowledge of genetics among medical doctors, especially those who come into contact with patients with genetic disorders for the first time. At the same time, the public understanding of human and medical genetics in Slovakia is inadequate, in spite of the fact that the general educational level in this country is comparable with that of most European countries. This will slowly improve as the national curriculum for secondary school education is now devoting more time to topics in genetics. At present there is no public discussion on ethical issues in the field of medical genetics in genetics. What is badly needed is a broad-scaled public education in the field of human genetics through all accessible educational means. None runs at present in Slovakia.

Historical and social background
Introduction

HENK TEN HAVE
Department of Ethics, Philosophy and History of Medicine
Catholic University of Nijmegen
The Netherlands

One of the reasons that genetic screening is continuously sensitizing the public and moral debate is the complicated history of human genetics. The scientific analysis of human heredity and the science of genetics in general are relative recent phenomena. From the start, practical applications have been imagined to improve humanity and to stop degeneration, prompting Galton to introduce the term 'eugenics'. In this section, Hans-Peter Kröner carefully dissects the history of human genetics, distinguishing three periods. The first period, starting with Galton's eugenics, focused on race, social class and selection. Ethics was subsumed under the normative rules of the biological sciences. Kröner describes how eugenics movements developed in many countries, with for example sterilisation laws in the US as well as the Scandinavian countries. The social darwinist orientation of this period degenerated into the National Socialist racial hygiene in Germany. The second period is characterised by the scientific explorations of molecular biochemistry. It started in the fifties, with the discovery of the molecular basis of human life, the development of new methods and techniques, and the elucidation of the chromosomal basis of diseases and disabilities. It is a period of transition and establishment of a new paradigm. The third period begins with the clinical applications of the molecular paradigm. In connection with new reproductive technologies, clinical genetics is able to influence human reproductive decisions and to apply screening techniques. Although the focus has shifted from population concerns to individual autonomy, human genetics, according to Kröner, is increasingly medicalising human life.

R. Chadwick et al. (eds.), The Ethics of Genetic Screening, 129–130.

Scepticism, critique and even distrust of human genetics is modelled on the historical example of abuse in Nazi Germany, the final stage of Kröner's first period. The Nazi experience is not only important in genetics, but also more generally in medical ethics. Although the experience is an important point of reference in most western countries, the impact is most thoroughly felt in Germany itself. Urban Wiesing analyses in the second chapter in this section the repercussions of history in present day debates in Germany. His argument is that historical abuse induces many to reject a clear-cut distinction between normality and abnormality, health and disease, sometimes even to deny the relevancy of this distinction for the moral debate. Philosophers, handicapped organisations and geneticists seem to agree that this distinction is not morally relevant. Wiesing also analyses the social responses to genetics in German society. Being one of the few countries with a significant anti-bioethics movement, open public debate on genetics is hampered in Germany. The historical experience imposes a special sensitivity on the public as well as special responsibilities on genetic scientists and clinical geneticists. The impact of history cannot be eliminated, as Wiesing argues, but there is no alternative to open and sincere debate on the ethics of genetics.

Use of genetic knowledge and technology is influenced by the historical background of human genetics. It is also determined by the social context in which genetic knowledge is obtained and genetic technology is applied.

Mairi Levitt in the third contribution to this section discusses the sociological perspective on genetic screening. The social context not only is the social and political structure in which screening programmes are developed, but also the value structure that determines the interactions between health providers and patients. A sociological approach therefore complements the standard ethical approach, because it draws attention away from the technical and individual level to the social and political context of genetics. Levitt argues that in particular three dimensions of the social context are in need of analysis. First, the introduction of screening programmes can be associated with social inequalities. Second, the genetic vocabulary focuses on the individual, shifting the locus of responsibility away from organisations and governments. Third, decision-making in real-life settings may be different from the ideological stance of non-directiveness in the application of genetic knowledge. Sociological research can help to clarify these dimensions of the social context.

As the contributions in this section show, history and sociology bring perspectives into the moral debate on genetic screening that focus attention on the wider context over time and in culture. These perspectives attribute to the moral evaluation of genetic screening programmes.

Chapter 12

From eugenics to genetic screening
Historical problems of human genetic applications

HANS-PETER KRÖNER
Institute of Theory and History
Munster University
Germany

1.   INTRODUCTION

When trying to deal with precursors of problems which modern human genetics or medical genetics confront us with, particularly in regard of genetical screening, some preliminary reservations must necessarily be made. Genetic screening as a relatively inexpensive method of testing larger populations quickly and technically simple for certain genetical traits, has just been thirty years old. Attempts at phenotypically registering carriers of definite physiological or pathological traits considered to be hereditary could be mentioned among the precursors of screening. But these methods were complicated and had been carried out with a different intention. It was only in the fifties, that the terms 'human genetics' and 'medical genetics' and their application in medicine were beginning to be accepted world-wide. The systematical scientific examination of human heredity, however, had started already in the second half of the 19th century. It was combined with reflections on the practical relevance of the new genetical knowledge. The application of genetics on human populations with the aim of preventing a degeneration or even improving the genotype was called eugenics by Galton. Still, the term 'eugenics' does not coincide with the modern notion of an applied science. Eugenics, as 'Eugenics Movement' in particular, has always been a technocratical movement intent upon solving social problems with scientific, biological means. Its followers saw themselves as prophets of a new, imminent positivistic religion of human betterment. But then

R. Chadwick et al. (eds.), The Ethics of Genetic Screening, 131–145.

again eugenics is part of the history of human genetics and medical genetics. Its scientific exponents were as a rule geneticists or doctors who often made important contributions to the science of human heredity.

The history of human genetics can roughly be divided into three periods. The first period, the period of 'classical eugenics' begins with Galton's establishment of the new science in the late 19th century and ends with the breakdown of National Socialism at the latest, when this kind of eugenics had become irrevocably discredited. The second period is marked by a growing medicalisation of human genetics, is rather a phase of passage and ends with the establishment of the molecular genetical paradigm in human genetics, this being the third period which still continues.

2.   THE FOUNDING OF CLASSICAL EUGENICS

Francis Galton, a cousin of Charles Darwin, had coined the word 'eugenics' in 1883, intending it to denote the science of improving human stock by giving "the more suitable races or strains of blood a better chance of prevailing speedily over the less suitable" (Galton, 1883, pp. 25-26). Galton demanded life-long imprisonment of habitual criminals, restrictions of the reproduction of the feeble-minded and mentally ill, a review of the benefits for the homeless and unemployed as to their possibly dysgenic effects, and the promotion of eugenically desirable marriages. (Pearson, Vol. 3). Realisation and public acceptance of this programme required a further expansion of genetical knowledge. Yet, what was even more important, 'national eugenics' had to become an 'orthodox creed' of the future; for up until now, the churches had practiced the arts 'a breeder would use who liked to breed cruel, vicious, and dull creatures'. It was a miracle that there had remained still enough good blood in the veins of the Europeans to reach the now rather moderate height of natural morals. (Galton, 1910) Galton's biographer, his disciple Karl Pearson, compared his master to a prophet from the old testament:

> harbinger of new morals, of an unfamiliar doctrine of altruism and like all new creeds difficult to be accepted and easy to be derided. Help more the strong one, this creed should read, support rather tomorrow's man than today's; see to it that knowledge and prudence control the blind emotions and rash instincts with which nature, with bloody teeth and claws, drives man, mindlessly and thoughtlessly, down her own evolutionary paths (Pearson, Vol. 3, p. 379).

The eugenicists demanded ethics to be in accordance with the findings of biological science. Until Galton, the term 'inheritance' was generally understood to denote a common tendency or force of like to beget like. Galton preferred the word 'heredity', which was not widely used then, and defined it as a physiological connection between the generations, which was open to research and quantification. Being a hereditary determinist he believed that essential human qualities like intelligence, vitality, vigour but also zeal, labour-loving instincts or tameness of disposition were inheritable (Soloway, 1990) By statistical examinations of larger populations over several generations he had found out that extraordinary qualities in a parent generation tended to regress to a mean in the descendant generations. Galton feared the 'British race' would come down to mediocrity, if one did not check this development by deliberate breeding, by an artificial selection of the 'best'. There was, however, an alarming observation that just the classes which were believed to be the most talented and destined to play a leading part in society tended to keep their families small, their birth rate thus no longer sufficing to guarantee their duration. The lower, less talented classes, on the other hand, multiplied excessively thus endangering national efficiency.

The early eugenicists were social Darwinists and believed that the social class system was the result of a process of selection and was therefore reflecting biological fitness. Christian morals of compassion and its more secular equivalents, the ideals of the French Revolution, democratic and socialist aspirations materialised in welfare state measures like benefits for the sick, the poor or the unemployed and finally the achievements of modern medicine and hygiene had created a social climate in which the struggle for life had been more and more abandoned, had even been conversed to its very contrary, the preferential selection of the 'inferior', the 'unfit'. Mass examinations of school children and recruits seemed to confirm this tendency.[1] The danger of degeneration - an objective one in the eyes of many contemporary scientists - could only be met by scientifically planning reproduction, by positive eugenics i.e. by promoting the reproduction of the 'desirable ones', and by negative eugenics i.e. by preventing the reproduction of the 'undesirable ones'. Such a programme necessarily clashed with the established values of individualist ethics. Eugenic ethics did not focus any longer on the individual but on a collective, and not even on a collective of living individuals but on a time-transcending abstraction namely the race, nation or people as a future ideal of perfection. The eugenicists knew that conflict only too well. Classical eugenics can be regarded as the very attempt to mediate between the requirements of science, of a rigorous Darwinism and the facts of society. Ploetz, for instance, the founding father of German eugenics or *Rassenhygiene*, a word he had coined, had subtitled his fundamental book,

'An essay on racial hygiene and its relation to the humanistic ideals, particularly to socialism' (Ploetz, 1895).

3.   FROM UTOPIA TO REALISATION

Ploetz had used a utopia to describe the effects of consequently applied eugenics on a society, 'the ideal process of race' as he called it.

> It is the outline of a sort of racial hygienic utopia, [he tried to soothe his readers], so the reader need not be frightened at its strange and cruel outward appearance as it is only a utopia from a one-sided, not exclusively legitimate point of view, which only serves to elucidate the conflict of the ideas of certain Darwinian circles with our cultural ideals (Ploetz, 1895, pp. 143-144).

Ploetz went on to describe an 'iatrocratical' society, a society i.e., in which a board of physicians decided on citizenship and the civic rights. At the end of an education which was to arouse a strong sense for racial well-being the young people should undergo a physical and mental examination, on which depended their right to marry and the number of children permitted to them. If unexpectedly a weak or deformed child was born, the physicians would let it have an 'easy death'. The same went for all twins and triplets and for all children who were born after the sixth sibling, after their mother's 45th or their father's fiftieth birthday. The economic struggle for life had to be completely maintained, but everybody should have the same chances: their was no law of succession in this society; private property went back to the state after its possessor had died. Benefits for the poor, the sick, and the unemployed should be reduced to a minimum. There was nothing to be said against wars as a means in the struggle for life among nations. But they had to be fought either by mercenary armies or it had at least to be ensured there was universal conscription regardless of the recruits' physical conditions:

> During the campaign it would be good to gather the particularly aligned bad variants at places where mainly cannon-fodder was needed and where individual fitness did not count so much (ibid., p. 147).

The eugenicists' preference for utopias cannot be explained only by the futurity of their subject, the generations to come, but is also owed to the insufficiency of their means i.e. the disproportion between knowledge and practical ability.[2] One of the problems of positive eugenics was which qualities were desirable from an evolutionary point of view. Biological fitness was a relative notion and depended on the prevailing conditions. The same applied to the biological common places, mentioned again and again, like strength, beauty, health, and intelligence. Eugenicists localised those qualities preferably in their own class, the middle class, as the traditional élites were considered to be biologically 'exhausted', whereas the lower classes were regarded as the reservoir of the less talented. Those opinions were not undisputed, though, as there was a strong socialist group in the eugenics movement, in Britain, for example, circling around Havelock Ellis and Caleb Saleeby, who considered such conceptions, particularly in regard of the working class, as partial. Consequently, there have been developed only few positive eugenic programs beyond some general plans for promoting large and healthy families. Even the Nazis realised only the beginnings of their formalistic aim of breeding a long haired, blond, and blue-eyed race.

Correspondingly, an attitude of defence prevailed in the eugenics movement, i.e. prevalent were reflections on how the danger of degeneration could be adequately met. So it is no wonder that negative eugenic measures in particular resulted in legal enactments. In Britain, eugenicists celebrated the 'Mental Deficiency Act' of 1913 as a victory of their movement. The law provided under certain circumstances the compulsory asylum of feeble-minded people, but also of poor people, of habitual drunkards and of unmarried pregnant women if they lived on state benefits for the poor. Yet the criterion for such a committal was not the hereditability of a person's condition but his social inability to care for his own life (Kevles, 1985). In Europe, eugenicists looked with envy at the U.S.A., where as early as in 1906 a law providing compulsory sterilisation was passed in Indiana. Fifteen further states followed until 1917. The laws provided the sterilisation of feeble-minded, habitual criminals, and sexual delinquents, provided they lived in a state institution and had thus become a public charge. The alteration of marriage laws is another legal measure, promulgated under the impact of the eugenic discussion in about thirty federal states by 1914. The altered laws declared null and void the marriages of the feeble-minded and of mental patients and provided a list of 'marriage obstacles' such as oligophrenia, alcoholism, communicable diseases etc. Finally the eugenicists executed a decisive influence upon the enactment of the Immigration Restriction Act in 1924. The law was supposed to restrict the immigration of people from east and south Europe and from non-European countries, as these people were regarded to be 'unamerican', 'non-adaptable', and 'racially inferior' (Ludmerer, 1972).

The sterilisation laws were not undisputed in the U.S.A, and critics kept on emphasising the low degree of certainty one had as to the hereditability of certain diseases according to the actual scientific knowledge. The problem was temporarily solved by a decision of the U.S. Supreme Court. In a case that went down in legal history as *Buck versus Bell*, the director of the 'Virginia Colony for Epileptics and Feebleminded', John H. Bell, had ordered the sterilisation of seventeen year old Carrie Buck, who thereupon went to law. The court confirmed the constitutional lawfulness of sterilisation. Justice Oliver Wendell Holmes who wrote the court's opinion pointed out:

> We have seen more than once that the public welfare may call upon the best citizens for their lives. It would be strange if it could not call upon those who already sap the strength of the State for these lesser sacrifices...in order to prevent our being swamped with incompetence. ...The principle that sustains compulsory vaccination is broad enough to cover cutting the Fallopian tubes. ...Three generations of imbeciles are enough (Kevles, 1985, p. 111).

Justice Holmes opinion only worded fears then wide-spread among educated circles. The *horror degenerationis*, the fear to be 'swamped' by the 'degenerated masses' reflected the social insecurity of large parts of the middle classes, who faced their actual social decline in the economic crises of the twenties. Still, the American model was only slowly imitated in Europe. In 1928, the Swiss canton Waadt passed a sterilisation law. Denmark followed in 1929. Other Scandinavian and some Baltic States passed sterilisation laws in the thirties. All these laws provided a eugenic sterilisation but on a strictly voluntary basis.[3] In Germany as well as in Britain, attempts of finding a majority for a sterilisation law failed in the twenties. Here too, the major part of the eugenicists endorsed voluntary sterilisation of the hereditarily ill. They opposed compulsory sterilisation, because they believed that such a massive interference in the rights of the individual could not be justified according to the actual state of their science, and what was more important, a public awareness of the necessity for eugenic action had not yet sufficiently evolved. Premature compulsory measures, however, they were afraid could turn out to be detrimental to the eugenic cause. So it is no wonder that in Germany under the conditions of dictatorship eugenicists were rash to give up their objections against compulsory sterilisation celebrating almost unanimously the *'Law for the Prevention of Hereditarily Ill Offspring' (Gesetz zur Verhütung erbkranken Nachwuchses)* of 1933. The law was based on a Prussian bill of 1932 and provided the registration and, in contrast to the Prussian bill, compulsory sterilisation of the hereditarily ill according to a list of indications. The number of people sterilised under the law is estimated to be about 400 000.

As their options for action were so limited, eugenicists directed their main attention on propaganda and public eugenic education. Only for that purpose, a 'Eugenic Education Society' had been founded in Britain, where prominent educated laymen like H.G. Wells or G.B. Shaw but also clergymen like Dean Inge of St. Paul's Cathedral fought for the eugenic cause. The guidelines (*Leitsätze*) of the 'German Society for Racial Hygiene', set up in 1922 and reformulated in 1931, culminated consequently in the demand for eugenic instruction and education, these being the indispensable precondition for reaching the final ends of racial hygiene. A decisive emphasis was to be laid upon the "renovation of the view of life (*Lebensanschauung*) towards a sense of eugenic responsibility". (Richtlinien, 1932, pp. 370-371). And Hitler's rather vague race hygienic programme, as set up in his book *Mein Kampf*, ended eventually in the call for new ethics. Physical and mental weakness is no shame, this ethics went, but it is a crime to burden innocent beings with that weakness for mere selfish reasons. But it would testify to greatest noble-mindedness, if a sick person renounced the right to have children of his own and preferred to adopt a healthy orphan instead (Hitler, 1931, pp. 446-448). With the racial hygienists, Hitler shared the naturalistic and anti-individualist creed. Individuals existed only as carriers of predominantly physical qualities and in relation to the greater, time-overspanning collective of 'race' or 'people' to which they owed the duty to have children or the sacrifice to renounce them. 'Duty' and 'sacrifice' were to become idealised virtues offered to the individuals as a miserable compensation for the fact that they had ceased to exist as autonomous, suffering and sensitive subjects. In Nazi racial hygiene a collectivist ethics of sacrifice - You are nothing, your people is all -, cost-benefit-calculations measuring the individual only by his economic worth for the whole, and a racism in a scientific, anthropological guise joined forces to wage a war of extermination against all who did not meet the National Socialist concept of race and achievement. The international community of eugenicists had partly welcomed with benevolence the National Socialist eugenic measures, particularly the ones in a stricter sense of the word such as the sterilisation law, or had at least assumed a wait-and-see attitude to observe the 'large-scale social experiment'. A growing protest, however, arose abroad against the anti-Semitic measures of the Nazis (Barkan, 1991), and what was more, population-genetical statistics had found out that it would take hundreds of years to eliminate detrimental genes in a given population by the traditional means of isolation or sterilisation. Even leading National Socialist human geneticists like Otmar von Verschuer knew about the population-genetical findings (Verschuer, 1939). At the same time, the findings of mutation research and statistical calculations gave the impression that the 'genetic load' weighed much heavier on mankind than one had supposed so far.

4.   FROM THE GENETICISTS' MANIFESTO TO POST-WAR HUMAN GENETICS

This 'change of paradigm' of genetics was the subject of the Seventh International Congress of Genetics, August 1939 in Edinburgh. The American 'Science Service', a scientific news agency, had turned to the congress with the question of how the world's population could be most effectively genetically improved. The answer titled 'Social Biology and Population Improvement' went down in the history of science as the 'geneticists' manifesto' (Social Biology, 1939). The manifesto was signed by 23 Anglo-American scientists, tending politically rather to the left, among them such renowned names as H.J. Muller, J.B.S. Haldane, L.T. Hogben, J.S. Huxley, J. Needham and T. Dobzhansky. It was obvious that one of the authors' main thrusts aimed at Nazi-eugenics though it did not explicitly say so in the text. The authors rejected a conception which ascribed a monopoly of good or bad genes to distinct social classes, nations or races and demanded first of all an improvement of social conditions as a basis for developing a social awareness and a sense of responsibility as to the procreation of eugenically desirable children. More research was to be invested in the production of more effective means of contraception, in the control of human fertility and the monthly period, and in artificial fertilisation. Voluntary sterilisation and abortion for eugenic reasons ought to be legalised. A deliberate control of selection should promote health, the so-called complex of intelligence and such dispositions as were more favourable for a sense of solidarity and the social behaviour than the qualities which stood for personal success today. If this programme was realised within a few generations, everybody could claim genius and strength as his birthright. Prerequisite would be a gigantic research effort on the field of human genetics and a co-operation of specialists from medicine, psychology, chemistry, and last not least the social sciences. The last end would have to be the improvement of the inner constitution of man.

Despite all concessions to the importance of nurture the manifesto remained rooted in biologistical, hereditary, and technocratical thought. It was Ploetz who had hoped as early as 1895 that the growing genetical and biological knowledge would permit to transfer selection to the level of the germ-cells. If one could successfully "separate the instant of procreation from the sensual urges of the moment, irresistible as they only too often would be, and postpone it to a desirable date of more favourable conditions", the conflict between the requirements of the struggle for life and the humanistic ideals would be solved (Ploetz, 1895, p. 235). The geneticists' manifesto too traced out a research program at the end of which stood the human genome project on the one hand, modern medicine of reproduction with in-vitro-fertilisation on the other.

When after the breakdown of National Socialism the terrible truth about the concentration camps and the murder of the mental patients, euphemistically called euthanasia, spread, classical eugenics seemed to be definitely discredited. Lancelot Hogben for instance, British geneticist, declared in 1945, much to the embarrassment of his eugenic colleagues, that "the official creed of the 3rd Reich permeated our own eugenics - from Galton to Dean Inge". C.P. Blacker, moderate British eugenicist, tried to shrug off this statement as something that had to be expected from a leftist like Hogben. According to Blacker, the Nazi-experiment in the concentration camps had absolutely nothing to do with eugenics as it was understood in this country. These words notwithstanding, Blacker wrote a few years later that in the light of the historic experience "personal liberty should [...] be specially treasured when demographic policies are devised" (Soloway, 1990, pp. 350-351). Most certainly, the end of National Socialism meant also the end of the eugenics movement. But that did not mean the end of eugenic thought. There are many reasons to believe that the eugenicists submerged so to speak, because eugenics which had always been a subject rather in educated circles, was now considered by those very circles to be 'politically not correct'.

It is no wonder therefore that eugenic thought had some kind of a revival when new possibilities of application began to show up on the field of human genetics. After classical eugenics with its rigorous measures had failed, human genetics in the fifties lacked options for action, as the new molecular genetics was not yet ripe for application. Medical application was limited to genetic counselling and its diagnostical and prognostical tools, all in all Mendelian examinations of family trees and statistical hereditary prognosis, did not differ very much from the ones used in the classical period.[4] Both in the U.S.A. and in Britain and West Germany, the process of institutionalisation of human genetics was promoted by national atomic research programs. James Neel from Ann Arbor, a member of the Atomic Bomb Casualty Commission, had examined the genetical after-effects on atomic bomb casualties of Hiroshima and Nagasaki (Kevles, 1985). It was generally known from Muller's radiation experiments in the twenties, that ionising radiation could cause mutations. The atomic programs of the fifties, both on the civilian and the military sector and here particularly atmospheric bomb-testing, gave reason for the fear the expected increase of environmental radioactivity might bring in its wake an increase of the human rate of mutation. To encounter this danger it seemed to be necessary to determine first of all the natural human rate of mutation as a starting point. 'Radiation Genetics and Mutation Research in Man' was one of the main topics of the First International Congress of Human Genetics in

Copenhagen in 1956. On an ensuing meeting of the WHO a team of scientists worked out a report on the effects of radioactive radiation upon the human genome with special recommendations for the promotion of human genetics. It was recommended there among other things to register hereditary defects and diseases in order to determine the natural rate of mutation. In West Germany, the registrations were sponsored by the federal ministry for atomic energy (Kröner, 1997).

In the second half of the fifties, new cyto-genetical methods can be attributed to the medicalisation of human genetics. After in 1955 Tjio in Sweden finally determined the correct number of human chromosomes as being 46, Lejeune found out in France in 1959 that Down's syndrome was caused by an additional chromosome. Almost at the same time in Britain, Turner's and Klinefelter's syndrome were recognised as variations of the normal number of chromosomes. The new optimism of progress which inspired the human geneticists also brought back old eugenic dreams of utopia. 27 prominent scientists, for instance, discussed the biological future of mankind on the Ciba-symposium in London in 1962 (Jungk, 1966). Almost all scientists shared the belief in Muller's thesis of the 'genetical load' which tended to be heavier and heavier due to the increase in radiation and the progress of civilisation. To meet this danger Julian Huxley demanded the development of new methods of influencing human reproduction such as oral contraceptives and multiple fertilisation by deep-frozen sperm from selected donors. Muller followed suit and called for 'eutelegenesis', for 'germinal choice' by 'AID' (artificial insemination from a donor). Sperm-banks were to be established to store the sperm of particularly talented men. The project was later realised as 'sperm-bank of the Nobel-prize-winners'. (Like all eugenicists these scientists too thought themselves to be the crown of creation and wanted to shape future generations according to their own image.) To Joshua Lederberg, molecular genetics promised the means of a direct genetical improvement of mankind "which would get us within only one or two generations of eugenic practice to a height to be reached today only within ten or hundred generations" (Jungk, 1966, p. 294). J.B.S. Haldane finally did not want to rule out that man could be adapted by genetic engineering to extreme environmental conditions originally alien to his nature. Legs, so he dreamt, would be useless for astronauts who had to travel for many years in space whereas a spider monkey's gripping tail would serve them much better. He also believed that higher mental qualities could be evolved by a proliferation of the brain. He played down the risks for the first subjects of such experiments by making them involuntary heroes of their 'pioneering' parents:

I conclude from the eagerness with which parents in our time encourage their children to risk their lives in a war that in a society with ideals different from ours many parents would be willing to risk the life of their little child hoping that it might develop extraordinary powers (Jungk, 1966, p. 388).

That this was not the swan song of a dying out species of eugenicists like Muller, Huxley or Haldane is shown by the fact that the cream of the young molecular geneticists like F. Crick, P.B. Medawar or J. Lederberg had not only been present but had taken an active part in the eugenic thought experiments.

5.   HUMAN GENETICS IN THE AGE OF MOLECULAR BIOLOGY

The sixties mark the beginning of the first programs of genetical screening on a larger scale. After Robert Guthrie had developed a simple method of testing new-born babies for phenylketonuria (PKU), a hereditary disease which could be treated by a special diet when detected in time, postnatal PKU-screenings were inaugurated in many American cities from 1966 on.[5] In New York City 51 PKU-children were identified between 1966 and 1974. It has been calculated subsequently that the cost of the screening program had amounted to one million dollars whereas the prospective cost for nursing the children in special institutions would have amounted to thirteen million dollars (Kevles, 1985). The cost-effectiveness of the PKU-screening led to the development of further postnatal screening methods such as galactosemia-screening. But their were also critics reproaching the exponents of PKU-screening of too much faith in technology as the Guthrie-test occasionally yielded false positive results or, when applied too early, false negative results. In both cases the consequences for the children would be dire: a phenylalanine deficient diet given to healthy children could lead to disability and a normal diet fed to PKU-children would result in mental retardation (ibid.).

A further step was the development of methods screening for heterozygous carriers of traits of recessive diseases. In order to prevent testing every potential parent couple, it was decided in the U.S.A. to test only populations holding a high risk for special diseases. These were particularly black Americans, who held a high risk for sickle cell anaemia, furthermore descendants from Mediterranean immigrants who suffered more often of thalassemia (Cooley's anaemia) and Ashkenasic Jews who were prone to Tay Sachs disease. Criticism against a compulsory screening, the way it had been established in a number of federal states, grew when some major American companies introduced genetical screening for prospective employees. When four black recruits suddenly died after a training in high altitude and the autopsy showed severe sickliness of the red blood cells the Air Force excluded black carriers of the sickle trait from the flying personnel and prohibited their access to the Air Force Academy. Spokesmen of the black community thereupon accused sickle cell screening as discriminatory, as a kind of eugenics against black people and even as a step towards genocide (ibid.).

Pre-natal screening by amniocentesis was eventually also developed in the late sixties. By the middle of the seventies it was possible to diagnose 'in utero' all chromosomal variations like Down's syndrome and 23 inborn errors of metabolism such as Tay-Sachs-disease. As a consequence a number of countries changed their abortion laws by introducing a eugenic or genetical indication.[6] Amniocentesis and legal abortion led to a boom in genetic counselling. In Germany human geneticists demanded an extension of pre-natal examination practice to a systematic screening of all elder prospective mothers and justified their demands with a cost-benefit analysis.[7] Criticism was advanced particularly by anti-abortionists. Another point of criticism referred to the status of handicapped people in society. There was fear that an animosity against handicapped people, widespread as it was anyhow, could be enhanced by a social pressure to have only 'healthy children'. In Germany the discussion was influenced by the memory of the Nazi-euthanasia crimes and by the theses of Australian philosopher Peter Singer. It was argued that denying a handicapped fetus the right to live would lead to denying living handicapped people the very right.[8] There was even criticism from the side of eugenics. Before the introduction of amniocentesis, the argument went, a couple with a Tay-Sachs-child as a rule would have decided not to have children anymore and thus would not have transferred their recessive disposition to their descendants. Amniocentesis, however, offered them the opportunity to select 'healthy' children, who might be heterozygous for the recessive gene, though. The gene was thus transferred to their descendancy and could slowly accumulate in the gene pool (ibid.).

Genetic mass examinations have accompanied the history of human genetics from its very beginning. Davenport's (famous American founder of eugenics) collection of genetical family dates in his Eugenics Records Office in Cold Spring Harbor was ostensibly meant to serve science alone but comprised as a rule socially stigmatised families. Mass examinations of so-called racially mixed populations, carried out for example by the anthropologist Eugen Fischer at the beginning of the century, were made with the outward intention to prove the validity of Mendel's laws for man but served only too often to confirm the prejudice that racial mixing was detrimental.[9] Large-scale genetical registration was conducted by the Nazis. They enacted a duty of notification for hereditary diseases, which was the prerequisite for sterilisation and later to some extent for 'euthanasia'.[10] Genetical registration, as it was carried out in the fifties with regard to a possible change in the human mutation rate by the increase of environmental radioactivity, was in my opinion also meant to calm a population worrying about the atomic plans of their respective governments. In this way, the social acceptance for the new 'weird' technology was to be enhanced.

It would be important to check whether such hidden motives also underlie today's projects of genetical screening. New human genetics in its medicalised form had claimed to serve only the actual individual in accordance with classical medical ethics and free from any eugenic collective utopias. Not the gene pool but the autonomy of the individual should be the guideline of genetic counselling. The British human geneticist Lionel Penrose was convinced that the autonomous and reasonable patients, and those were the majority, were willing to avoid serious risks and to accept moderate ones, and he predicted that:

> the result of skilful counselling over a long period of years, will undoubtedly be to diminish, very slightly but progressively, the amount of severe hereditary diseases in the population (Kevles, 1985, p. 258).

His American colleague James Neel, however, stated in 1971:

> Any population policy - or for that matter, no population policy - may have implications more far-reaching for the gene pool than all the genetic counselling of the next hundred years (ibid., p. 258).

The discussion about socio-biology and the revival of theories of a genetical determination of intelligence and behaviour, reflections at last on the relationship between individual and community in a society with limited resources suggest that questions are being publicly discussed again, which had been considered for a while as being historically incriminated, as not opportune. 'Consequently, it follows', Richard A. Soloway writes in his book about eugenics in Britain:

> that if eugenic demography was submerged by political and social change rather than permanently defeated in the scientific and social scientific arena and therefore remains latent in modern culture, it would not be surprising to see it re-emerge in a different political and socio-economic environment, perhaps under a different name, sanctified by

a modern science that is itself in large part a product of that same culture and environment (Soloway, 1990, p. 362).

NOTES

1 The protractedness of the Boer War had been explained among other things with the alleged bad physical condition of the British soldiers. The Inspector General of recruiting had reported in Parliament that eight out of eleven volunteers in Manchester had to be rejected as physically unfit (Kevles, 1985, p. 73.).

2 Not only Ploetz had invented a utopia but also Galton, which had been declined by his publisher on the ground that it might hurt the moral sense of the Victorian readers. (Soloway, 1990, pp. 66-67) There are other eugenic utopias e.g. Haldane, 1925, or Muller, 1935.

3 If a person had no or restricted legal capacity, however, the legal representative or guardian could decide on the sterilisation of his ward.

4 That does not mean that there has not been any progress of knowledge in the second phase: I mention only Neel's discovery of sickle cell anaemia being a recessive hereditary disease and Linus Pauling's biochemical confirmation of this discovery as being a haemoglobinopathy.

5 Phenylketonuria (PKU) can be treated by a phenylalanine-deficient diet, when diagnosed in time. Otherwise the disease will result in severe mental retardation.

6 Great Britain 1967, U.S.A. 1973. In Germany too the law provided a 'eugenic indication', though it was rather a social one as it did not refer to the gene pool but was meant to prevent a situation which might become unbearable for the parents. The term was later changed to infantile or embryopathic indication and is now, as the new abortion law of 1995 provides, part of the maternal indication, i.e. the decisive question is now whether it can be expected from the mother to continue pregnancy and to raise a handicapped child afterwards.

7 The authors calculated, taking Down's syndrome as a model, that 48 Million DM could be annually saved by systematical pre-natal screening of all pregnant women being 38 or older. That sum would otherwise have to be raised for the nursing and medical care of Down's patients. The study was awarded the prestigious 'Hufeland-prize' of the German Board of Physicians in 1977 (Passarge, 1979).

8 Cf. e.g. Hegselmann, 1991, or Christoph, 1990.

9 Cf. e.g. Fischer, 1913.

10 As to the part 'hereditary statistics' played in the '3rd Reich' cf. Aly, 1984.

11 Soloway believes, however, that in Britain at least the social climate is not very inviting for a eugenic revival, (Soloway, 1990, p. 362).

REFERENCES

Aly, G. et al., 1984. *Die restlose Erfassung - Volkszählen, Identifizieren, Aussondern im Nationalsozialismus,* Rotbuch Verlag, Berlin.

Barkan, E., 1991. *The retreat of scientific racism - Changing concepts of race in Britain and the United States between the World Wars,* Cambridge University Press, Cambridge.

Christoph, F., 1990. *Tödlicher Zeitgeist - Notwehr gegen Euthanasie.* Kiepenheuer and Witsch, Köln.

Fischer, E., 1913. *Die Rehobother Bastarde und das Bastardisierungsproblem beim Menschen,* Gustav Fischer Verlag, Jena.

Galton, F., 1883. *Inquiries Into Human Faculty And Its Development,* Macmillan, London.

Galton, F., 1910. *Genie und Vererbung,* Leipzig. (1869, *Hereditary Genius - An Inquiry into its Laws and Consequences,* Macmillan, London). Quotations are from the German edition and retranslated into English by me.

Haldane, J.B.S., 1925. *Daedalus oder Wissenschaft und Zukunft,* Drei Masken Verlag, München.

Hegselmann, R. et al., 1991. *Zur Debatte über Euthanasie,* Suhrkamp, Frankfurt.

Hitler, A., 1931. *Mein Kampf,* (7th edn), Eher Verlag, München.

Jungk, R. et al. (ed.), 1966. *Das umstrittene Experiment, der Mensch,* Verlag Kurt Desch, München Wien Basel. (English Edition, Wolstenholme, G. (ed), 1963, *Man and his Future.* Boston Toronto.) Quotations are from the German edition and retranslated into English by me.

Kevles, D. J., 1985. *In the Name of Eugenics - Genetics and the Uses of Heredity,* Alfred A. Knopf, New York

Kröner, H.-P., 1997. Förderung der Genetik und Humangenetik in der Bundesrepublik durch das Ministerium für Atomfragen in den fünfziger Jahren, in, K. Weisemann et al. (ed), *Wissenschaft und Politik - Genetik und Humangenetik 1949-1989,* Lit-Verlag, Münster, Hamburg, London, forthcoming.

Ludmerer, K., 1972. 'Genetics, Eugenics, and the Immigration Restriction Act', *Bull. Hist. Med.,* 46, pp. 59-81.

Muller, H.J., 1935. *Out of the Night - A Biologists View of the Future,* Victor Gollancz, New York.

Passarge, E. et al., 1979. *Genetische Pränataldiagnostik als Aufgabe der Präventivmedizin - Ein Erfahrungsbericht mit Kosten/Nutzen-Analyse,* Ferdinand Enke Verlag, Stuttgart.

Pearson, K., 1914-1930. *The Life, Letters and Labours of Francis Galton,* (4 vols), Cambridge University Press, Cambridge.

Ploetz, A., *1895. Die Tüchtigkeit unserer Rasse und der Schutz der Schwachen,* S. Fischer Verlag, Berlin.

'Richtlinien der Deutschen Gesellschaft für Rassenhygiene (Eugenik)', 1932 (Guide-lines of the German Society for Racial Hygiene [Eugenics]), *Archiv für Rassen- und Gesellschaftsbiologie,* 26, pp. 370-371.

'Social Biology and Population Improvement', 1939. *Nature,* 144, pp. 521-522.

Soloway, R. A., 1990. *Demography and Degeneration, Eugenics and the Declining Birthrate in Twentieth-Century Britain,* Chapel Hill, London.

Verschuer, O. v., 1939. 'Bemerkungen zur Genanalyse beim Menschen', *Der Erbarzt,* 7, pp. 65-69.

Chapter 13

Genetics in Germany
History and hysteria

URBAN WIESING
Department of Medical Ethics
Eberhard-Karls University, Tübingen
Germany

1.   INTRODUCTION

In Germany more than any other country, references to historical misuse of medicine feed the medical-ethical discussion.. Of course, many discussions in Germany are hardly to be distinguished from others taking place in the western world. However, some very typical German reactions towards new technologies in general and genetics in particular must be reported. And without any doubt these reactions are rooted in the German history and sometimes tend towards a hysterical refusal of anything related to genetics. While the discussion always refers to the German history, the question arises as to whether the typical German attitude towards genetics, and even towards bioethics as a whole, is the only lesson to learn from Nazi medicine.

This chapter tries to answer the question by analysing the reply given in Germany to two questions closely related to genetics: What is discussed about the delineation between normality and abnormality, and what are the reactions to genetics in general and genetic screening in particular? The results may serve as a basis for discussing the central question: Are the typical German reactions justified considering the history?

R. Chadwick et al. (eds.), The Ethics of Genetic Screening, 147–156.

2.   NORMALITY OR ABNORMALITY?

What is discussed about the delineation between normality and abnormality? This question cannot be answered generally for Germany, because too many discourses with different approaches and different opinions take place. They take hardly any or no notice of each other and refer to the German history differently. Therefore I will distinguish between the discourse among theorists, especially philosophers, among authors of the 'cripple movement', as it came to be called, and among geneticists to present a rough idea of the prominent lines of thought.

The philosophical discourse - especially in post-modern philosophy - and the discourse of medical theory emphasises the uselessness of the delineation between normality and abnormality - no normality for man in the world is given and plurality and variety have not only to be tolerated but to be enforced (Welsch, 1991). Especially the normative aspect of normality and abnormality is denied and any mandatory connection between normal/abnormal and health/disease is regarded to be unjustifiable and even dangerous (Honnefelder, 1996). Some medical theorists emphasise that health and disease are deontological terms, which means they require action in a therapeutic sense, and they cannot exclusively be derived from sciences and from any kind of normality (Wieland, 1985). The patient's subjective feeling must necessarily be a part of a definition of health and disease. Karl Eduard Rothschuh (1972, 1975) mentioned that the main criterion for a definition of disease should be the subject requiring help. Therefore a delineation between normality and abnormality is not necessary. That means: theorists and philosophers are sceptical about the benefits of the delineation between normal/abnormal. They refer to the Nazi medicine at most indirectly: They oppose any attempt to use scientific biological knowledge in a normative sense - which was a common strategy of Nazi medicine.

The organisations of handicapped people and the authors of 'cripple movement' reject any idea dealing with normality/abnormality very vigorously and they use the historical example of Nazi medicine frequently. They emphasise the right to abnormality and suggest that abnormality can even be creative, enriching or a positive challenge (Neuer-Miebach and Tarneden, 1994). In an inquiry among 27 *Selbsthilfegruppen* (self-help groups) most of their representatives stated that "life with a handicapped child has a different but not necessarily a poorer quality" (Liese, Zerres 1991, p. 56).[1] This idea has to be promoted among experts in particular: "Physicians have to learn that being handicapped is a special kind of health" (Kölner Manifest, 1994, pp. 148-149). And a popular slogan of the handicapped movement and others emphasises another conception of 'normal'; "It is normal to be different." (Verband für Anthroposophische Heilpädagogik et al., 1995, p. 155).

The 'Code of Basic Ethical Assumptions' of four major organisations engaged in the help for the handicapped focuses on normality as well: "Being handicapped is a kind of existence that cannot be distinguished from so-called normality" (Verband für Anthroposophische Heilpädagogik et al., 1995, p. 155).[2] Of course being handicapped means being abnormal according to a statistic idea of normality, but this statistic approach fails completely. If the idea of normality includes any normative aspects it will be dangerous for all human beings, not only for the handicapped: "The idea of a human being made to measure is directed against all of us." (Kölner Manifest, 1994, p. 149) Oliver Tolmein, one of the most prominent journalists writing against the new technologies and against a debate about eugenics and euthanasia[3] speaks of "the terror of normality" (Tolmein, 1989, p. 27). In addition to this danger authors closely related to the 'cripple movement' are afraid that genetics even enforces the question of normality and abnormality: "Genetechnology intensifies the force to straighten life according to standards. If genes and biogenetic processes are feasible it will have to be defined which processes should be accepted and which should not." (Neubeck-Fischer, 1991, p. 65).

The question of the delineation between normality and abnormality is hardly discussed among practitioners in genetics and consensus has not yet been achieved and published.[4] Only Propping states (without mentioning the Nazi era): "From the genetic point of view 'normality' can not be determined, rather in human genetics variety is the 'normality'." (1996, p. 106) The impression is that the question for the practitioners arises whether and in which way the idea of normality or abnormality has a function for the aim of the therapy - knowing that abnormality is definitely not always enriching and creative, but sometimes painful for the subject and its relatives. Most people feel healthy when their body is to a certain extent close to normality. Normality has no moral aspect but is, at least in a heuristic sense and when weighted carefully, helpful for intervention. For genetic practice it is hardly possible to work without the idea of normality or abnormality. However, the strategy sometimes suggested to use other words such as *auffällig/unauffällig* (conspicuous/inconspicuous) is a linguistic avoidance strategy and not very convincing.

3.   THE SOCIAL RESPONSE TO GENETICS

What is the social response to genetics in the German society? On a scientific level no basic objections to genetic screening are raised. But besides a wide discussion on the point a very German phenomenon can be observed: a comparatively sceptical attitude of major parts of the population as well as some political parties towards anything dealing with genetics and the so-called anti-bioethics movement. This movement was initiated on the occasion of two planned lectures by Peter Singer in 1989 by the handicapped and by the 'cripple movement', but it quickly gained support from several left oriented groups, 'alternative groups', feminist groups, some religious groups and some journalists and scientists. Although the movement started on the occasion of Singers lectures on euthanasia the issues to be attacked rapidly widened to genetics, genetic engineering, reproductive medicine and even bioethics itself and its intellectual background, analytical philosophy and utilitarianism.[5] The members of the anti-bioethics movement consider the anglo-saxonian bioethics and its German version to be "deadly ethics" (Bruns et al., 1990), the "ethics of killing" (pamphlet in Sass, 1990, p. 45) and "a public relation department of the new biotechnologies" (Klee, 1990, p. 81).[6] There are some relations between the anti-bioethics movement and the anti-nuclear power movement - personally and regarding the sceptical attitude towards industrialisation.

The anti-bioethics movement tries to prevent a public debate about the issues mentioned before because they believe that immoral and dangerous views are made respectable only by being discussed. Some militant members do not hesitate to restrict tolerance and to use violence to achieve their aims; they even feel obliged to do so according slippery slope arguments. "All measures aimed at selection must be rejected without compromise! Tolerance against ideas, discussions and actions that radically question the life of many human beings are not allowed to exist." (Anonymous author, in Sass, 1990, p. 42-43). The members vigorously state that all these issues are part of a development that leads us once again to the Nazi atrocities.

These critics in Germany believe genetic counselling to be the prolongation and direct follower of the eugenics during the Nazi era, as the sub-title of a book wants us to believe: '*Vom Erbgesundheitsgericht zur humangenetischen Beratung*' (From the court of genetic health [a Nazi court] to genetic counselling) (Sierck and Radtke, 1984). Therefore the opponents demand the closure of any centre offering genetic counselling: "We demand to close all institutes of genetic counselling and centres for pre-natal diagnostic and to abandon the life-threatening diagnostic of selection!" (Anti-Euthanasie-Forum, published in Neuer-Miebach and Tarneden, 1994, p. 150-151)

Considering that this movement contains only a few groups and a very few journalists it is very effective. Some examples: the annual conference of the 'European Society for Philosophy of Medicine and Health Care' in 1990 had to be cancelled a week before its beginning in Bochum and had to be moved to Maastricht because of enormous agitation and announced riots. Several university seminars about ethics were disrupted or could not take place as well as the 15th of the traditional Wittgenstein-Symposia in Kirchberg 1991, which topic was to be applied ethics; a committee to fill a professorship in applied ethics in Hamburg had to negotiate at a secret place because of announced disruptions; and it is hardly possible to let Peter Singer or Helga Kuhse speak in Germany. Their book *Should the Baby Live* was to be published in a German translation in the highly reputed Rohwolt Verlag. After protests the publisher changed his mind; it is now published in the Harald Fischer Verlag. This list of examples could be prolonged.[7]

Closely related to this small but effective 'anti-bioethics movement' a sceptical and critical attitude towards anything concerning 'genes' can be recognised in major parts of the population and especially in the green party (Bündnis 90/Die Grünen) which came out third in the last election. In a study about genetic engineering and newspapers the following results were presented - and most probably the same is right for genetics, genetic engineering is mainly discussed in the political sections of the newspapers and less often in the scientific sections; and in the political section genetic engineering is viewed mainly sceptically and negatively, in the scientific sections mainly positively (Kepplinger, 1995). Also the German philosophers and theologians are partly sceptical about genetic screening because of the amount of unsolved problems. And according to the principle that problems have to be solved first and then a new measure has to be established Dietmar Mieth, a well known theologian and ethicist, pleads for a preliminary stop: "In spite of all advantages that can be described and cannot be denied a fundamental approval of genetic counselling and pre-natal diagnostic appears to me to be early and questionable as long as the problems of genetic counselling and the informed decision are only realised and not solved" (Mieth, 1991, p. 136).

The reason for the anti-bioethics movement and for the sceptical attitude in some parts of the population can only be found in German history and this explains among other things the success of the anti-bioethics movement.[8] The most influential weekly newspaper, *Die Zeit*, asked, "Is it possible to speak in Germany about human genetics without thinking of Auschwitz?" (Albrecht and Rückert, 1994, p. 20) which is certainly the central question. For many people any dealing with genetics or euthanasia and any discussion about these topics is dangerous and has to be suppressed, because they consider the total rejection to be the only way to prevent us from a catastrophy similar to that of the Nazi era. And the counter-arguments of the geneticists - namely that they are working with a totally different approach and trying to implement nondirective counselling concepts in order to prevent this catastrophy from happening again - these arguments cannot override the cautious attitude in parts of the population and do not convince the members of the 'anti-bioethics movement' at all.

On the other hand in the last few years the willingness of the self-help groups and the geneticists to discuss the crucial issues of genetics together is increasing. An *Arbeitskreis Selbsthilfegruppen und Humangenetiker* (Working group of self-help groups and human geneticists) was founded in 1993 with the aim to promote the exchange of opinions and achieve guidelines acceptable for both.

German history influences the German geneticists as well despite the fact that they are the target of the 'anti-bioethics movement'. An inquiry found out that a majority of German geneticists consider themselves to be ethically more sensitive and more restrictive in genetic screening programs than their international colleagues. They complain about the fact that genetics in Germany is always confronted with and blamed for what happened 50 years ago and that genetics can not be discussed adequately in their country.[9] Because of the Third Reich's experience German geneticists think that they have a special responsibility for a person at risk to have children with genetic disease and a person affected with genetic disease. Compared with the geneticists in Portugal and the UK German geneticists "were significantly more likely to counsel in favour of keeping affected pregnancies" (Marteau et al., 1994, p. 98).

Germany's historical burden influences the politicians as well. The then Minister of Law, Mrs. Leutheuser-Schnarrenberger, confirmed on the occasion of the rejection of the European Bioethics Convention[10] that Germany has due to its history a 'special responsibility' in bioethics. But - and not surprisingly for a politician - she did not explain what 'special responsibility' means.

4.   THE PARADOX

The statement of the former Minister of Law raises the question which meaningful lesson could be learned from Nazi medicine? Furthermore, the German reactions suggest us to clarify the role that a historically given mentality should play in a moral debate. The shadow of the Nazi trauma on the German reactions towards genetics cannot be eliminated. Generally spoken attitudes towards demanding contemporary problems are necessarily influenced by the history of a country. Therefore the cautious attitude towards genetics in Germany is unavoidable as well as understandable. And this attitude could even be fruitful in the international debate about genetics. Fruitful in a double sense: It could make us aware of crucial developments and it could remind us of the role of historically given mentality and emotions, as Nicholson suggests: "The greatest value that could come from the present protests [in Germany] would be if it were to make bioethicists like Singer sit down to some creative thinking on how to move the subject away from a sterile rationality to a process that takes full account of human nature and emotions" (Nicholson, 1990, p. 23).

But should a mentality be respected that oppresses academic freedom and rejects any discussion about certain issues? Is the total rejection the only way to prevent us from a catastrophy similar to that of the Nazi era? The demand to stop discussions about genetics is a paradox lesson for several reasons: It is paradox because it prevents us from learning Nazi medicine. The implication of the Nazi era for contemporary medicine and its bioethical problems "cannot be discussed if we refuse to discuss bioethics" (Wikler and Barondes, 1993, p. 52). Additionally, among other things the Nazi outrages were made possible by a totally oppressed public debate. And it is indeed an "irony ... that the protesters have themselves shown the kind of fanaticism and lack of respect for rational debate that was also a necessary pre-condition of the Nazi atrocities" (Singer, 1990, p. 42). Considering these pre-conditions the sometimes hysterical refusal to discuss these issues is not understandable and it is a paradox that in Germany "often the same authors who vehemently criticise modern medicine's inhumane vitalism also attack as presumptuous and dangerous any attempt to regulate these problems in a generalised way" (Schöne-Seifert and Rippe, 1991, p. 27).

There is no reasonable alternative to an open, engaged and sincere debate about the ethical issues in genetics. The German historical experience even supports this way of dealing with the demanding problems related to that new technology. Many participants of the public German debate about genetics should reconsider the lessons of the holocaust.

NOTES

1 All translations by Urban Wiesing.
2 The code seems not to be scrutinised by its implications. The organisations demand for example: "Experiments with human life must be forbidden." (p. 155) This demand is neither realistic nor morally acceptable. A well-founded consensus has been established in Germany as well as in other countries that the experiments with human beings (both clinical and non-clinical) are morally sound under certain conditions. These conditions are listed up in numerous declarations, most prominently in the revised Declaration of Helsinki. Sentences like the last quoted show that the declaration of the four organisations is not always written thoroughly.
3 Kottow (1988) confirms that the debate about euthanasia is also deeply influenced by the German history: "...the trauma of holothanasia has left long-living scars and deformations in the public discussion of euthanasia in Germany. ...In the foreseeable future and possibly for a long time to come, pro-euthanasia arguments will be nipped in the bud by pointing an accusatory and perhaps overzealous finger at associations with the not too distant past." (Kottow, 1988, p. 67).
4 This result is confirmed by Prof. Wolff, Freiburg.
5 Peter Singer gave a comment in Bioethics (Singer, 1990): "Freedom of speech that is unable to challenge social consensus is not worthy of the name." (p. 42) The reason for the vigorous opposition to his ideas can be found - according Singer - in a special tradition of reasoning: "There appears to be a near-total lack of tradition of rational argument in practical ethics in Germany." (p. 43). The influence of the 'Singer Affair' on bioethics in Germany in general is described by Leist 1993.
6 See also Schöne-Seifert and Rippe 1991.
7 The events following the invitation of Peter Singer are well documented; see for example Anstötz 1990, Anstötz et al. 1995, Schöne-Seifert and Rippe 1991, Hegselmann and Merkel 1991, Bastian 1990, Nicholson 1990.
8 Schöne-Seifert and Rippe (1991) provide further arguments for the success of the anti-bioethics movement.
9 Personal information given by Prof. Imgard Nippert, the co-ordinator; the project is funded by the Deutsche Forschungsgemeinschaft and NIH.
10 More about the reaction to the proposal of the European Bioethics Convention in Germany see Bockenheimer-Lucius (1995).

REFERENCES

Albrecht, J. and Rückert, S., 1994, 'Das natürliche Schicksal'. *Die Zeit*, 3 November, pp. 17-20.
Anstötz, Ch., 1990, 'Peter Singer und die Pädagogik für Behinderte', *Analyse & Kritik,* 12, pp. 131-148.

Anstötz, Ch., Hegselmann, R. and Kliemt, H., (eds), 1995. *Peter Singer in Deutschland. Zur Gefährdung der Diskussionsfreiheit in der Wissenschaft*, Peter Lang, Frankfurt a.M.

Bastian, T., 1990. 'Wie dieses Buch entstand... Vorbemerkungen des Herausgebers', in T. Bastian (ed.) *Denken - Schreiben- Töten. Zur neuen 'Euthanasie'-Diskussion*, Hirzel, Stuttgart, pp. 9-14.

Bockenheimer-Lucius, G., 1995. 'Die 'Bioethik-Konvention' - Entwicklung und gegenwärtiger Stand der Kontroverse', *Ethik in der Medizin* 7, pp. 146-153.

Bruns, T., Pensellin, U. and Sieck, U. (eds), 1990. *Tödliche Ethik*, Verlag Libertäre Assoziation, Hamburg.

Hegselmann, R. and Merkel, R. (eds), 1991. *Zur Debatte über Euthanasie. Beiträge und Stellungnahmen*, Suhrkamp, Frankfurt a.M.

Honnefelder, L., 1996. 'Humangenetik und Pränataldiagnostik. Die Normative Funktion des Krankheits- und Behinderungsbegriffs: Ethische Aspekte', in L. Honnefelder and C.F. Gethmann (eds) *Jahrbuch für Wissenschaft und Ethik*, (Vol. 1), Walter de Gruyter, Berlin New York, pp. 121-128.

Kottow, M.H., 1988. 'Euthanasia after the Holocaust - Is it Possible? A Report from the Federal Republic of Germany', *Bioethics*, 2, pp. 58-69.

Kepplinger, H.M., 1995. 'Die Gentechnik in der Medienberichterstattung', in W. Barz, B. Brinckmann and H.J. Ewers (eds) *Gentechnologie in Deutschland.Umweltschutz, Gesundheitsschutz, Wirtschaftsfaktor, Akzeptanz*. Lit Verlag, Münster, Hamburg, pp. 195-214.

Klee, E., 1990. *'Durch Zyankali erlöst': Sterbehilfe und Euthanasie heute'*, Sterbehilfe und Euthanasie heute, Fischer, Frankfurt a.M.

Kölner Manifest, 1994. 'Vom Recht auf Anderssein', Published in: Th., Neuer-Miebach and R. Tarneden, (eds), *Vom Recht auf Anderssein. Anfragen an pränatal Diagnostik und humangenetische Beratung*, Lebenshilfe Verlag, Verlag Selbstbestimmtes Leben, Marburg, Düsseldorf, pp. 148-149.

Leist, A., 1993. 'Bioethics in a Low Key. A Report from Germany', *Bioethics*, 7, pp. 271-279.

Liese, P. and Zerres, K., 1991. Selbsthilfegruppen und Humangenetik - Ergebnisse einer Umfrage von Vertretern von 27 Selbsthilfegruppen, in K. Zerres and R. Rüdel (eds) *Selbsthilfegruppen und Humangenetik im Dialog*, Ferdinand Enke Verlag, Stuttgart, pp. 52-57.

Marteau, T., Harriet, D., Reid, M. *et al.*, 1994. 'Counselling following Diagnosis of Fetal Abnormality: A Comparison between German, Portuguese and UK Geneticists', *European Journal of Human Genetics*, 2, pp. 96-102.

Mieth, D., 1991. 'Genomanalyse - Pränataldiagnostik, ethische Grenzen?', in D. Beckmann, K. Istel, M. Leipoldt and H. Reichert (eds) *Humangenetik - Segen für die Menschheit oder unkalkulierbares Risiko?* Peter Lang, Frankfurt a.M., Bern, New York, Paris, pp. 125-142.

Neubeck-Fischer, H., 1991. 'Gentechnologie. Acht Thesen zu ihren gesellschaftlichen Voraussetzungen und Folgen', in D. Beckmann, K. Istel, M. Leipoldt and H. Reichert (eds) *Humangenetik - Segen für die Menschheit oder unkalkulierbares Risiko?*, Peter Lang, Frankfurt a.M., Bern, New York, Paris, pp. 61-70.

Neuer-Miebach, Th. and Tarneden, R., (eds), 1994. *Vom Recht auf Anderssein. Anfragen an pränatale Diagnostik und humangenetische Beratung*, Lebenshilfe Verlag, Verlag Selbstbestimmtes Leben, Marburg, Düsseldorf.

Nicholson, R., 1990. 'Bioethics attacked in Germany', *Bulletin of Medical Ethics*, pp. 19-23.

Propping, P., 1996. 'Humangenetik und Pränataldiagnostik. Die Normative Funktion des Krankheits und Behinderungsbegriffs: Medizinisch-humangenetische Aspekte', in L.

Honnefelder and C.F. Gethmann (eds), *Jahrbuch für Wissenschaft und Ethik,* (Vol. 1), Walter de Gruyter, Berlin, New York, pp. 105-110.

Rothschuh, K.E., 1972. 'Der Krankheitsbegriff (was ist Krankheit?)', *Hippokrates,* 43, pp. 3-17.

Rothschuh, K.E., (ed.) 1975. *Was ist Krankheit? Erscheinung, Erklärung, Sinngebung,* Darmstadt, Wissenschaftliche Buchgesellschaft.

Sass, H.M. and Viefhues, H., 1990. *Vierte Jahrestagung der European Society for Philosophy of Medicine and Health Care.* Programm, Abstracts, Dokumentation, 59, Medizinethische Materialien, Bochum.

Schöne-Seifert, B. and Rippe, K.P., 1991. 'Silencing the Singer: Antibioethics in Germany', *Hastings Center Report,* 21, 6, pp. 20-27.

Sierck, U. and Radtke, N., 1984. *Die Wohltäter-Mafia. Vom Erbgesundheitsgericht zur Humangenetischen Beratung,* Selbstverlag, Hamburg.

Singer, P., 1990. 'Bioethics and academic freedom', *Bioethics,* 4, pp. 33-44.

Tolmein, O., 1989. 'Terror der Normalität', *Konkret,* 7, p. 27

Verband für Anthroposophische Heilpädagogik, Sozialtherapie und Soziale Arbeit e.V., Verband Evangelischer Einrichtungen für Menschen mit geistiger und seelischer Behinderung e.V., Verband Katholischer Einrichtungen für Lern- und Geistigbehinderte e.V. and Bundesvereinigung Lebenhilfe für geistig Behinderte e.V., 1995. 'Ethische Grundaussagen der vier Fachverbände der Behindertenhilfe', *Ethik in der Medizin,* 7, pp. 154-155.

Welsch, W., 1991. *Unsere postmoderne Moderne.* Weinheim, Acta humaniora, (3rd edn).

Wieland, W., 1985. 'Grundlagen der Krankheitsbetrachtung', in R. Gross (ed), *Geistige Grundlagen der Medizin,* Springer Verlag, Berlin, Heidelberg, New York, Tokyo, pp. 42-55.

Wikler, D. and Barondess, J., 1993. 'Bioethics and Anti-Bioethics in Light of Nazi Medicine: What Must We Remember?', *Kennedy Institute of Ethics Journal,* 3, pp. 39-55.

Chapter 14

A sociological perspective on genetic screening

MAIRI LEVITT
Centre for Professional Ethics
University of Central Lancashire
England

1.   INTRODUCTION

Social science has an acknowledged role in carrying out empirical research on the effects of screening programmes which provides data for medical professionals and resource managers. However, it will be argued that sociology can contribute to discussion in ethics not simply by the provision of empirical data but also by the different theoretical frameworks it employs which lead to different perspectives on the same topics as those which engage philosophy and psychology.

The relationship between the individual and society is basic to sociological study. Sociology examines the social influences on behaviour, values and identity. The area of genetics seems particularly apt for a study of the broader sociological effects as well as individual psychological effects because genes are shared and the diagnosis of genetic disorders has implications for other family members. Genetic technology also raises questions for society as a whole, which arise from programmes brought in in a piecemeal fashion.[1] The question of which conditions to screen for and when termination of pregnancy, preimplantation or other measures to avoid the birth of a child with the disorder should be employed, are questions both for the individual and society. While unease is expressed in general about reproductive technology, in specific cases public and media sympathy appears to lie with the individual against regulation, as in the Diane Blood

R. Chadwick et al. (eds.), The Ethics of Genetic Screening, 157–166.

case in England.[2] Diane Blood had asked for semen to be taken from her husband's body while he was in a coma suffering from bacterial meningitis and thus unable to give his consent. He later died and she requested permission to have artificial insemination using her dead husband's sperm. There is a reluctance to interfere with a specific individual's choice. The 'ethos of late capitalism' has been described as "the world of options, lifestyles and preferences" (Bruce, 1996, p.233). With the decline in active participation in the Christian churches in Europe most individuals select beliefs and moral choices, not in terms of tradition, but by deciding what is best for themselves. The individualist language of counselling and screening thus appeals to a core value in most European societies but should not be left unexamined.

Discussion of the ethical implications of genetic screening has been led by philosophers who consider such issues as the costs and benefits of screening, moral responsibility and the requirements of non-maleficence and beneficence (Shickle and Chadwick, 1994). In general sociologists will focus on the social context in which screening programmes are developed, both at the macro and micro level. At the macro-level the economic and political structure will constrain choices and shape the programmes. For example, knowledge of existing structural inequalities in access to power points to areas for research into inequalities of provision and treatment in the health service. At the micro-level of interaction and taken-for-granted values and attitudes, sociologists might focus on the language and terminology used between medical staff and patients. While geneticists and other medical professionals are the 'reality definers' their version of what is going on cannot be accepted as 'reality' until it is investigated and compared with versions of those less often heard; the patients, potential patients and their families. There is no prior assumption that the accounts of medical staff are more valid than others.

2.   PUBLIC UNDERSTANDING OF GENETICS

Public concern about genetics and developments in genetic technology is still often portrayed as due to ignorance, with the implication that the dispelling of ignorance will allay concern. As Layton *et al* discuss in relation to families with Down's syndrome babies, scientific understanding of the condition has little relevance to the practical knowledge which parents need and acquire for themselves (Layton, et al., 1993). Whether or not the public are 'ignorant' will depend on what they are being asked.

Ignorance of technical terms and the techniques of genetics are no doubt widespread but in a report on research using focus groups this lack of knowledge was "no bar to sophisticated discussion of the ethical and social issues raised" (Vines, 1997, p.1055). A survey of young people age 11 to 18 in four European countries including Britain, asked children to respond to stories about genetics including one on xenotransplants (without using the term itself). The children raised the same social, ethical and practical concerns about the technique as those found in the Nuffield Council of Bioethics report (Levitt and Whitelegg, 1996, p.132f; Nuffield Council, 1996). Not only were they unlikely to have been scientifically literate, they also came from countries where there is, as yet, little media interest in genetics. A sociological approach would be to ask how a concentration on technical knowledge serves to direct the debate on genetic technology away from the social and political context in which the technology will be used (Barns, 1996). Thus while one approach would be to look at genetic screening as a matter of solving technical problems of diagnosis and opposition to screening as pointing to the need for more medical/scientific information, another would be to look at the particular social context in which the technology is employed. Once it is in use, technology cannot be 'uncontaminated' by that context and 'information' is not a neutral tool but will have been prepared and presented by individuals with their own interests and values. Lippman and Wilfond consider written material on genetic conditions and argue that "we must recognise that every description of a genetic disorder is a story that contains a message [and] no single story, however balanced, can ever be neutral or value free" (1992, p.936).

Lippman uses the term 'geneticization' to describe the 'dominant discourse' in discussion of health and disease; "genetics is increasingly identified as *the* way to reveal and explain health and disease, normality and abnormality" (Lippman, 1994, p.144). This opens the way for the medicalisation (or re-medicalisation) of conditions which have been seen to have social and psychological components; aggression, alcoholism and intelligence (Mestel, 1993; New Scientist, 1997; Goddhart, 1994). Criticism of an overemphasis on genes as determining human behaviour exists within the scientific community but does not redress the imbalance; "...in relation to human behaviour...*most* geneticists will *probably* agree that nurture still has a place"(my emphases) (Maddox, 1993, p.107). A sociological approach would consider the possible social consequences of an acceptance of a genetic basis for complex conditions which could lead to radically different treatment for 'sufferers' and their families but no solution to the problems they cause. Even in simple gene disorders where the genetic basis has long been understood, such as sickle cell anaemia, it is not necessarily possible to predict the effect in any one individual, nor did the genetic information lead to the treatment now given. Mapping the genetic element in more complex conditions is likely to be less useful. For explaining the numbers meeting violent deaths in USA compared with the UK, the availability of guns is more useful than genes. However, reiterating the well known environmental and social factors influencing crime and the often expensive action to reduce it, will not provide media interest, research grants or opportunities for commercial interest. In contrast, research on the genetic factors offers possibilities of developing gene therapy and, more lucratively, drugs to treat the condition. There would be more commercial interest in developing a drug to counteract the effects of polluted water than in ensuring the water was clean in the first place.

In the following sections three aspects of the social context of screening will be examined; social inequalities, individualisation and decision-making. In some instances the social consequences of specific programmes have been well documented, for example, sickle cell testing in the USA. However, as will be shown, this does not mean that the social context has necessarily been included in the planning or carrying out of subsequent programmes.

3.   SOCIAL INEQUALITIES AND SCREENING PROGRAMMES

The consequences of a failure to look beyond the technical aspects of screening programmes can be seen in the following two examples which show the effects of gender inequality. A programme to identify carriers of Maple Syrup Urine (MSU) among a Bedouin tribe resulted in potential male carriers seeking wives outside the tribe while the women had to remain unmarried and dependent on their fathers (Carmi, 1991). An understanding of the differing power and status of men and women in that culture would have enabled the medical staff to predict, and protect women against, the devastating social consequences. Once the problem had been identified such action was taken. A letter from staff at a department of medical genetics in Lucknow, India, raised problems of genetic counselling of couples who were both carriers of a genetic disease. In the two cases cited harassment and a demand for second marriage by the husband occurred or was expected to occur (Naveed, Phadke, Sharma and Agarwal, 1992). Therefore it is necessary to include gender as a variable when considering the effect of carrier status. Given women's role in the family it is likely that the woman, rather than the man, will bear the greater responsibility for any long-term commitment required by a child born with a genetic disorder.

Research also suggests that women feel the burden of the genetic disorder more and feel more guilt (Somer, Mustonen and Norio, 1988).

Whilst it might be suggested by medical professionals that genetic screening techniques could remove some of the inequalities given by nature, a social perspective would lead to the (testable) expectation that the effects of screening procedures on individuals will mirror the inequalities in society; socio-economic class and racial inequalities as well as those of gender. If genetic therapy is limited to those who can afford to pay, then rich individuals/countries will have new choices available to them and the desired biological attributes will be unevenly distributed (Holm, 1994). Research in Finland has already shown that those having genetic counselling had a higher educational level than the general population and that those with a higher level of general education had more knowledge of the mode of inheritance and their family's risk for the relevant disorder (Somer, Mustonen and Norio, 1988). Knowledge of, and access to the new technology will be unevenly distributed as will the ability to cope with its consequences, for example any effects on the family's insurance status.

Despite the debate on the ethical and social implications of genetic research and screening, discussion over research into genes and IQ reveals familiar divisions between some of the geneticists, who are "not worried by the ethics" and "want to get on with the science" and ethicists and social scientists working in the area concerned with the way the findings will be used (Vines, 1996; Goodey, 1996, p.14). If intelligence is found to be largely determined by genetic inheritance then a differentiated education system would be the obvious choice, educating children according to their aptitude. As Goodhart put it in *New Scientist*, "If the differences really are inborn, there is no use forcing children beyond their natural capabilities"(Goodhart, 1994, p.51). Inevitably, different schooling would not mean equal schooling in terms of resources and status. The practical implications of such a policy in reinforcing inequalities rather than attempting to alleviate them are obvious.

4. INDIVIDUALISATION

A genetic explanation of disease and behaviour focuses on the individual (Lippman, 1994). A critical perspective would lead to the question of who benefits from genetic explanations, as opposed to those which focus on environmental and social causes? Leaving the decision to the individual absolves others of responsibility for its consequences. Once the problem becomes individual, protection is provided not only for the medical professional but for other interest groups. There have been cases of women employees in the USA losing their jobs after workplace screening found a high level of toxic chemical or compounds that are teratogenic (US Congress, Office of Technology Assessment, 1990). Focusing on the individual element of disease, rather than on aspects which would require social action and expenditure, is attractive to organisations and governments who might otherwise have to act to remove health hazards from the workplace or control pollution.

Advances in genetic applications have been hailed uncritically in the past but the right *not* to know is surfacing in media reporting (Chadwick and Levitt, 1997). A consideration of prostrate cancer screening resulted in a decision by the National Health Service against the introduction of routine screening because the costs (impotence, incontinence, postoperative mortality and psychological disturbance) outweighed the benefits (Stewart-Brown and Farmer, 1997). Criticism of this decision came from Woolf who saw the issue as one of individual empowerment; 'men over 50 have a right to decide for themselves about screening', the right to know operates 'regardless of whether the National Health Service funds it or whether patients will (or can) act on the information..' (Woolf, 1997). This perspective appears to regard decisions as a matter for informed individuals but Woolf goes on to suggest that shared decision-making should be reserved 'for situations when the superiority of one option is uncertain' . However, for the individual surely the superiority of one option is always uncertain since calculations are made on the basis of many cases. The individual might be the one in a thousand or ten thousand for whom the 'superior option' was a personal tragedy. It is also useful to consider the consequences of such a policy applied across society. Some will have information but no choice, if they rely on the NHS for health care, others will pay for repeated screens to check everything is all right.

Where screening is available handicap may be seen as avoidable. It has been suggested that genetic screening will encourage more intolerance of those who are less than perfect (McCarrick, 1993). The issues have already been raised in the USA in relation to 'wrongful life' cases and health insurance provided by employers (Clarke et al., 1992). Any restrictions of health care for those who refused genetic counselling and/or the abortion of affected fetuses would lead to further inequalities by religion, ethnic group and social class. Restrictions on the availability of some treatments in the UK have already been discussed and in some cases implemented; smokers being denied operations on the National Health Service. Policies could be pursued under the ethical value of individual responsibility and freedom of decision-making to which Western societies subscribe.

5. DECISION-MAKING

Angus Clarke's article entitled *Is non-directive counselling possible?* began with the statement "ostensibly non-directive counselling in connection with pre-natal diagnosis is inevitably a sham" (Clarke, 1991, p.998).   He evoked a critical response in the pages of the *Lancet*. One letter quoted the Royal College of Physicians' report (1989):

> The goal of genetic and pre-natal diagnostic provision must be to help these couples make an informed choice, one which they feel is best for themselves and their families (*Lancet*, 1991, p.1267)

Such a statement reiterates the official view of genetic counselling but begs the question, raised by Clarke, of whether it is attainable in practice.

Studies of attitudes frequently fail to probe beneath the surface of professional ideology to investigate practice.   Those who have undergone professional training might be expected to present themselves as practising its precepts.   Thus genetic counsellors trained in the United States, Canada or the UK will tend to present themselves as 'non-directive' when asked how they would proceed in hypothetical cases (Wertz and Fletcher, 1988). However, sociologists would display a degree of scepticism to a reiteration of the values inculcated in professional training by the 'reality definers' until these could be compared with everyday practice.   In education few teachers would fail to subscribe to the ethic of equality of opportunity for all children.   Rather than assuming the ethic is put into practice sociologists carry out empirical studies of everyday practice in schools.   When considering the practice of genetic screening, the views of those who have undergone counselling would be an essential part of any empirical study. A comparison of patient expectations with the counsellor's view of their role would be one area of investigation.   Patients may expect the professional to give them clear guidance from his or her special knowledge and expertise. Do patients attempt to find out what the counsellor really thinks they should do from whatever information is given?   No genetic test is 100 per cent accurate nor can the effects of a genetic disorder be precisely forecast (e.g. Fragile X syndrome; Webb 1993) but such uncertainty is problematic both for medical staff and patients.

Research which analysed transcripts of real counselling sessions found, in contrast to Wertz and Fletcher that counsellors did tell clients what was best for them (average 5.8 times per session) and whether they had made the 'right' decision (average 1.7 times per session). Lower social classes were most likely to be given opinions (Michie, Bron, Bobrow et al., 1997 p.44). The editorial points out that clients often seek the opinion of the counsellor. This is likely to be true but is not relevant to the issue of why counsellors seek to present themselves as non-directive. Research looking at specific tests found that despite this emphasis on individual choice and informed consent the most important variable in take-up rates for testing is the way it is offered (Bekker et al., 1993).

Another area for investigation is the way clients use the information they are given on genetic disorders. In a survey on attitudes to carrier screening among recent parents, the couples were given detailed information on the genetic disorders considered in the survey and it was assumed that they would take the information into account when answering questions (Green, 1992). Without an investigation of the decision-making process in real-life settings it cannot be assumed that the patient's understandings of what is going on are the same as the counsellors, nor that the information they are given will form the main basis of their decision. Questions to be raised would include; whether patients/parents are taking the same factors into consideration as medical staff/counsellors and the relative weight given to the advice of doctors, nurses, counsellors, partners and other relatives and friends. Richards discusses areas in which there may be a gap between the aims of those seeking genetic screening and the providers of the service and between medical and lay ideas of inheritance and risk (Richards, 1993). An understanding of 'lay' beliefs would assist in the assessment of future demand for screening services.

6.   CONCLUSION

As knowledge of the human genome increases the questions on the social implications of that knowledge become more urgent. The basic sociological approach is to question received wisdom, look at the point of view of all those involved and study what is actually going on. Sociological perspectives look critically at the new technology and ask how it will be used, whom it will benefit and what the wider social consequences will be. The questionnaire, which is favoured in studies evaluating reactions to a screening programme, is only one methodological tool and is less well suited to revealing the processes leading up to the formation of opinions and decisions or the underlying world-view of those involved. Digging more deeply makes the findings more complicated and less clear-cut so will be harder to simply label a programme as a 'success' or 'failure'. To concentrate only on take-up rates for screening programmes, the accuracy of results and the effects of results on individuals is to understand only part of the story.

NOTES

[1] Until the late 1980s parents in Britain were rarely offered any pre-natal testing other than for Down's Syndrome. Around 100 inherited conditions can now be screened for.
[2] The Human Fertilisation and Embryology Act bans the use of sperm in Britain without the donor's written consent. In 1997 the Court of Appeal ruled that she could have artificial insemination in another European Community country unless there were strong public policy reasons against (British Medical Journal, 1997. 314, p.461).

REFERENCES

Bekker, H., Modell, M., Denniss, G., Silver, A., Mathew, C., Bobrow, M., Marteau, T., 1993. Uptake of cystic fibrosis testing in primary care. Supply push or demand? *British Medical Journal*, 306, pp. 1584-6.
Barns, I., 1996. 'Manufacturing consensus. Reflections on the UK National Consensus Conference on Plant Biotechnology' *Science as Culture*, 5.2, 23, pp. 199-216.
Bruce, S., 1996. *Religion in the Modern World from cathedrals to cults,* Oxford University Press.
Carmi, R., 1991. 'Genetic counselling in a traditional society', Letter in *Lancet,* 337, pp. 306.
Chadwick, R. and Levitt, M., 1997. 'Mass media and public discussion in bioethics' R. Chadwick, M. Levitt and D. Shickle (eds), *The Right to Know and the Right Not to Know,* Avebury, Aldershot, pp. 79-86.
Clarke, A., 1991. 'Is non-directive counselling possible?' *Lancet,* 338, pp. 998-1001.
Clarke, A., 1992. 'Children with genetic diseases who should pay?' *Lancet,* 339. pp. 1614-1615.
Green, J.M., 1992. 'Principles and practicalities of carrier screening. Attitudes of recent parents', *Journal of Medical Genetics,* 29, pp. 313-319.
Goodey, C., 1996. 'Genetic markers for intelligence' *Bulletin of Medical Ethics,* pp. 13-16.
Goodhart, C.,1994. 'Does a brighter future beckon', *New Scientist,* 144, 1948, pp. 50-51.
Gordon, M., 1997. 'Boozing flies expose 'alcoholic genes', *New Scientist,* 154, 2078, p. 20

Holm, S., 1994. 'Genetic Engineering and the North-South Divide' in A. Dyson and J. Harris (eds), *Ethics and Biotechnology*, Routledge, London, pp. 47-63.

Layton, D., Jenkins, E., MacGill, S., Davey, A., 1993. *Inarticulate science? Perspectives on the public understanding of science and some implications for science education*, Studies in Science Education Ltd, Driffield.

Levitt, M. and Whitelegg, M., 1996. 'Empirical Research Findings' in R. Chadwick, H. Häyry, M. Häyry, M. Levitt, J. Lunshof and M. Whitelegg, *BIOCULT. Cultural and social objections to Biotechnology. Analysis of the arguments with special reference to the views of young people*, Report of a project funded by the Commission of the European Communities under the BIOTECH programme, Centre for Professional Ethics, University of Central Lancashire, Preston, pp. 118-159.

Lippman, A. and Wilfond, B.S., 1992. 'Twice-told Tales. Stories about Genetic Disorders', *American Journal of Human Genetics*, 51, pp. 936-937.

Lippman, A., 1994. 'Pre-natal genetic testing and screening. Constructing needs and reinforcing inequalities', in A. Clarke (ed), *Genetic Counselling. Practice and Principles*. Routledge, London, pp. 142-186.

Maddox, J., 1993. 'Has nature overwhelmed nurture?', *Nature*, 336, pp. 107.

McCarrick, P.M., 1993. 'Genetic Testing and Genetic Screening', Scope Note 22, *Kennedy Institute of Ethics Journal*, 3, 3, pp. 333-354.

Mestel, R., 1994. 'Does the 'aggressive gene' lurk in a Dutch family?', *New Scientist*, 141, 1914, pp. 31-34.

Michie, S., Bron, F., Bobrow, M. and Marteau, T., 1997. 'Nondirectiveness in Genetic Counselling. An Empirical Study', *American Journal of Human Genetics*, 60, pp. 40-47.

Naveed, M., Phadke S., Sharma, A. and Agarwal, S., 1992. 'Sociocultural problems in genetic counselling', *Journal of Medical Genetics*, 29, pp. 140.

Nuffield Council on Bioethics, 1996. *Animal-to-Human Transplants. The ethics of xenotransplantation*, Nuffield Council on Bioethics, London.

Pembrey, M., 1991. 'Non-directive genetic counselling', Letter in *Lancet*, p.1267.

Richards, M.P.M., 1993. 'The new genetics. some issues for social scientists', *Sociology of Health and Illness*, 15, 5, pp. 567-586.

Shickle, D. and Chadwick, R., 1994. 'The ethics of screening. Is 'screeningitis' an incurable disease?', *Journal of Medical Ethics*, 20, pp. 12-18.

Somer, M., Mustonen, H and Norio, R., 1988. 'Evaluation of genetic counselling. Recall of information, post-counselling reproduction, and attitude of the counsellees', *Clinical Genetics*, 34, pp. 352-365.

Stewart-Brown, S. and Farmer, A.., 1997. 'Screening could seriously damage your health', Letter in *British Medical Journal*, 314, 22 February, p.533.

U.S. Congress, Office of Technology Assessment *Genetic monitoring and screening in the workplace*, OTA-BA-455, U.S. Government Printing Office, Washington, D.C., p.148.

Vines, G., 1996. 'The Search for the Clever Stuff', *Guardian Newspaper*, G2 section, 1 February, p. 2.

Vines, G., 1997. 'Genetics. let the public decide', *British Medical Journal*, 314, pp. 1055.

Webb, J., 1993. 'A fragile case for screening?', *New Scientist*, 25 December, pp. 10-11.

Wertz, D.C. and Fletcher, J.C., 1988. 'Attitudes of Genetic Counsellors. A Multinational Survey', *American Journal of Human Genetics*, 42, pp. 592-600.

Woolf, S., 1997, 'Should we screen for prostate cancer? Men over 50 have a right to decide for themselves', *British Medical Journal*, 314, pp. 989-990.

Moral and philosophical issues
Introduction

Ruth Chadwick and Urban Wiesing
Centre for Professional Ethics
University of Central Lancashire
United Kingdom
Department of Medical Genetics
Eberhard-Karls University, Tübingen
Germany

In thinking about philosophical and ethical approaches to genetic screening, one question that arises is whether and to what extent consensus is possible in European bioethics. That leads to another question - what is meant by the term 'consensus'. In this introduction we first look at consensus in bioethics itself, and then consensus in practical application in practice. This discussion has implications for the way forward in bioethics.

If we begin by understanding consensus in terms of 'agreement', what is the focus of agreement? In bioethics consensus could mean agreement on specific issues or agreement on theoretical ethical frameworks. It seems fairly clear that agreement on specific issues in bioethics across Europe is unlikely, given the fact of moral pluralism and cultural diversity. Even within one society there are opposing views on issues such as euthanasia, abortion and gene therapy. At the level of theory an example of disagreement is the debate between the four-principles approach and its critics. These principles, of autonomy, beneficence, non-maleficence and justice, sometimes referred to as the 'Georgetown mantra', are defended on the grounds that they do form the basis of a consensus in that they can be agreed upon form a variety of theoretical positions (cf. Beauchamp and Childress, 1994). It has, however, been suggested that principlism is essentially American, and that the four principles may not travel well to other cultures (Holm, 1995).

Certainly the set of principles identified by the Group of Advisers to the European Commission as appropriate for ethical assessment of research in

R. Chadwick et al. (eds.), The Ethics of Genetic Screening, 167–170.

the life sciences is much wider than the four principles approach (Group of Advisers, 1998). They include (with respect to human beings): autonomy, respect for human dignity, non-discrimination, proportionality, non-exploitation and protection of the vulnerable.

The four principles are in considerable use in European bioethics, but not only do they have rivals such as casuistry and virtue ethics, they (and the items on the European list) are also susceptible to varied interpretations. So if there does exist in Europe a large degree of consensus on the value of autonomy, for example, what does that amount to if there is disagreement on what autonomy means?

There are two respects in which consensus in bioethics may have an important part to play. The first is agreement about the type of activity bioethics is, including the importance of rational dialogue itself, in bioethics, which is subject to challenge from the anti-bioethics movement. The second is the recognition of existing conceptual and ethical frameworks, certainly, but combined with an acceptance of the fact that they are constantly subject to reexamination and possibly reinterpretation in the light of new developments. The papers in this section explore the ways in which accepted concepts need to be examined afresh. Ingmar Pörn's paper, for example, looks at the concepts of care and health; Judit Sandor addresses the new angles that genetic screening offers to the concept of privacy.

Ingmar Pörn's paper, 'Genetic information and care' takes as a premise that genetic screening can only be morally justified if it meets acceptable care goals and proceeds to examine the extent to which genetic screening can be seen as an essential part of care, even though it is not a treatment as conventionally understood. The argument depends on the interpretation of health not as freedom from disease but in terms of a system of adequacy.

There has been much discussion about the implications of genetic information for confidentiality, privacy, the moral acceptability of disclosure and the right not to know. Sandor's paper points out a special problem for privacy in the genetic context, however, namely the issue of intergenerational privacy. Genetic information concerning future people may be stored even before they are born. How can privacy deal with this?

In addition to consensus in bioethics, there is the issue of consensus in society. The construction of a social consensus is an ongoing process involving dialogue between different sections of society. Dolores Dooley sets the debate in political context, showing how discussions about genetic services in the Republic of Ireland have been shaped by ongoing negotiations between values in tension: individual liberty and the common good. In the context of a strong theological interpretation of the common good, there has been a public silence about developments such as amniocentesis which might facilitate individual choice. Dooley warns that minimisation of public debate works to give power to a few, and that it needs to be surmounted in order to give due recognition to the reality of value pluralism.

Joseph Fletcher, in his examination of the possibility of consensus, has pointed to the possibility of consensus within a professional group (e.g. clinical geneticists) in reaching agreement on the types of ethical problems they face and in developing optimal approaches to these problems (Fletcher, 1994). The case of genetics and especially genetics screening , however, shows a phenomenon well known to other fields of contemporary medicine and bioethics. Simply spoken the issue is much more complicated than it looks on first sight. Despite consensus on a certain level of moral consideration in bioethics no consensus can be derived directly on a practical level in medicine. To understand this phenomenon several reasons have to be mentioned.

1. The consensus about the general moral aims of medicine cannot be transferred simply to the practical level. Different practical guidelines in genetic screening can be defended with reference to the same moral principles, as Rogeer Hoedemaekers points out in his article. To give an example: to help the sufferer is a well accepted moral principle in medicine. But is the knowledge about a certain genetic status really a help for the sufferer or not? What is to be done if the result of the test can be both - helpful and harmful - and only statistical probabilities can be given before testing?

2. The competition of different aims in medicine has always been a problem in bioethics. In genetic screening it is one as well: the autonomous decision to get knowledge about a certain gene may lead to harmful results for the patient and his or her relatives. To what extent is a physician allowed to refuse a test knowing that the result might be harmful and referring to the principle 'do no harm'? Sometimes genetic knowledge is knowledge not only about one person but about relatives and offspring as well. Therefore the question arises, whose autonomy and whose harm is at stake and how should these be balanced in case of conflict.

3. Certain moral questions in medicine cannot be answered consensually even on a theoretical level, In these cases it is much more difficult to get practical answers. For example the question of abortion - or morally speaking the question of the moral status of a fetus - is a highly controversial one. But the moral issue of pre-natal screening is largely dependent on the question whether abortion is morally sound or not. Any ethical consideration about pre-natal screening has to keep in mind the nowadays unavoidable plurality of thought about abortion.

4. In a complex situation one new factor can change the whole moral issue. A measure which is very acceptable for adults can be

unacceptable for children, as Angus Clarke shows us convincingly. Therefore each measure has to be scrutinised carefully and separately. No factor should be left out of consideration, because it could change the situation completely.

That means: the application of accepted moral principles in medicine is difficult, especially when the technical conditions require deep knowledge of the issue. A well grounded acquaintance and the power of judgement is needed. Any attempt to provide clear-cut and simple resolutions is most probable to fail. Therefore one task of moral philosophy could be to warn of the deceitful hope for a simple moral solution in the complex cases like genetic screening. One the contrary: moral philosophy should encourage going into the technical details but without losing the theoretical ethical background. If an example of the need for multidisciplinary work was needed, the case of genetic screening would provide it. It shows again how practical ethics has to proceed if it does not want to become futile.

REFERENCES

Beauchamp, T.L. and Childress, J.F., 1994. *Principles of Biomedical Ethics,* Oxford, University Press.

Fletcher, J.C., 1994. 'Can there be consensus on ethics in human genetics?' in *Ethics and Human Genetics*: Proceedings of the 2nd Symposium of the Council of Europe on Bioethics, Strasbourg, Council of Europe.

Group of Advisers [to the European Commission] on the Ethical Implications of Biotechnology, 1998. 'The ethical aspects of the Fifth Research Framework Program', *Politics and the Life Sciences,* 17, 1, pp. 73-6.

Holm, S., 1995. 'Not just autonomy - the principles of American biomedical ethics'. *Journal of Medical Ethics,* 21, 6, pp. 332-8.

Chapter 15

Genetic information and care

INGMAR PÖRN
Department of Philosophy
University of Helsinki
Finland

According to Gelehrter and Collins (1990, p. 270), screening tests "are an essential part of standard medical care and are generally directed at early diagnosis of treatable diseases". They go on to say:

> Sometimes screening tests and diagnostic tests are the same; more frequently, however, the screening test is aimed at identifying a subset of the population in whom more specific diagnostic tests should be performed. (loc. cit.).

They distinguish between two major types of genetic screening tests. "The first is aimed at the early recognition of affected individuals in whom medical intervention will have a beneficial effect for the affected individuals and/or the patient's family" (op. cit., p. 271). This type covers fetal screening, which includes pre-natal diagnostic tests and newborn screening. "The second major form of genetic screening is the identification of individuals at risk of transmitting a genetic disorder" (loc. cit.). They add, with some reservation for the hazards involved, including ethical issues, that population screening programs "may be carried out to obtain epidemiologic information about genetic diseases rather than to identify populations in whom medical intervention is indicated" (op. cit., p. 275).

In their discussion of criteria for cost effective programs for carrier screening Gelehrter and Collins stress the importance of the following factors: that a high risk population can be identified, that the disease is clinically significant and severe enough to warrant such a program, that an inexpensive test is available with adequate sensitivity and specificity, and

R. Chadwick et al. (eds.), The Ethics of Genetic Screening, 171–180.

that definitive tests are available for specific diagnosis in individuals identified as being at high risk.

In summary, Gelehrter and Collins, authors of an influential text in the field, see genetic screening as an essential part of standard medical care. Its role is that of being a step preparatory to more specific diagnostic tests aimed at the treatment and prevention of disease. And they stress that "prevention of genetic disease by genetic screening, counselling, pre-natal diagnosis, and early intervention is closely intertwined with treatment" (op. cit., p. 283).

The report by Nuffield Council on Bioethics is also concerned with genetic screening as part of medical care, and for this it lays down three goals (1993, p. 17):

1. contribute to improving the health of persons who suffer from genetic disorders, or
2. allow carriers for a given abnormal gene to make informed choices regarding reproduction, or
3. move towards alleviating the anxieties of families and communities faced with the prospect of serious genetic disease.

These goals are consistent with the characterisation of genetic screening given by Gelehrter and Collins, and although there are variations on the theme, they constitute a representative set of goals for genetic screening as part of medical care. As such they are of considerable interest. There can be no moral justification for genetic screening unless it meets acceptable care goals and does so in a reliable way. The errors allowed must result from marginal mistakes in the application of methods which yield beneficial effects most of the time. The identification of the genetic determinants cannot be allowed to become an end in itself; its importance must always turn on its role in informing treatment and prevention.

In this perspective the example of sickle cell anaemia, often used in attempts to justify genetic screening, is interesting, for it shows that "it is not necessary to know how genes work or where they are located to promote health and well-being, to understand illness, to relieve suffering" (Lippman, p. 1471), the reason being that "astute clinical observations and inferences, not genetic information, led to the demonstration that prophylactic penicillin would dramatically improve the quality of life of newborns with sickle cell anaemia" (loc. cit.).

Although the three goals are thought of as goals for genetic screening as part of medical care, their restriction to medical care will not be made in this paper. For an open investigation into the main forms of care constituting the setting for the use of genetic information it seems preferable to think of them quite generally as goals for the use of genetic information in care, not necessarily medical care exclusively.

In goal one above, health care is clearly the form of care intended. But how should the concept of health be articulated? According to an answer that has received some attention, health is the absence of disease, and disease is a reduction of functional capacity below statistically defined normal efficiency (for details, see Boorse, 1977). This conception is open to a serious objection: statistically defined normalcy is meaningful only in relation to a reference class; the possibility that a disease is statistically normal in the reference class, and hence not a disease, can be excluded only by assuming that the members of the reference class are normally healthy; but then circularity ensues since the concept to be articulated, namely health, is presupposed (cf. Wulff et al., 1990, p. 48).

A more promising approach is to give, first, a positive and disease-independent characterisation of health, and then go on to characterise disease and similar notions in terms of internal states and processes which have a causal tendency to undermine or reduce health. There are various ways of doing this (see for example Whitbeck, 1981; Pörn, 1984, 1993; and Nordenfelt, 1987). A criticism that may be passed on these approaches is that they do not pay sufficient attention to the value-laden character of the construction advanced. The criticism is justified, and it must be met, to sharpen the contrast between value-free articulations and those that in fact incorporate dependency on values.

According to the approach favoured in this paper, health should be articulated in relation to a system of adequacy. Its main features are as follows:

1. human beings are acting subjects;
2. the relations of the acting subject to her environment and to herself can be defined in terms of goal-directed action;
3. the main determinants of goal-directed action are:
   i. the agent's *repertoire* of abilities;
   ii. the agent's *environment* of external and internal circumstances; and
   iii. the *goals* of the agent, organised in her life plan (which need not cover the duration of her life).

The agent's state of adequacy at a given time is her system of adequacy at this time. This system is a complex relation between the three main determinants. Before we define this relation we attend to some features of its terms. We take them in the order just given.

Let us consider some descriptions that might apply to what a person did on a specific occasion: "He made a telephone call", "He made an international telephone call", "He telephoned his girlfriend", "He spoke to his girlfriend for six hours over the telephone". The list could be made very long indeed. What he did on that occasion can be described in different ways and at varying levels of generality. So when we speak of generic descriptions of actions we allow for variations in respect of generality.

Abilities, like actions, are specified generically at varying levels of generality. A's ability to X is relative to certain stated or implied conditions - these constitute the *range* of his ability. For example, the ability to swim is relative to certain conditions concerning distance, the height of the waves, water temperature, and so on. When nothing is said explicitly about the range there is a tendency to assume that the implied conditions are normal (Morriss, 1987). A's ability to X in conditions or circumstances C is present when A's Xing is compatible with the state of A at the time concerned and with C. We understand this compatibility to entail the conditional that if A chooses to X in C she will X in C.

Generic abilities are normally decomposable into abilities, in much the same way as goals may be subdivided into sub-goals. For example, the ability to give genetic counselling to a family in which there is a risk of a genetic disorder contains as a component the ability to assess the risk; and this in turn subdivides into the abilities to identify, estimate, and evaluate the risk. Generic abilities may also contain items of propositional knowledge as ingredients. The ability to make an international telephone call involves knowing the order in which the codes have to be dialled. And, importantly, the decomposition of generic abilities *always* gives physiological functions or functions at a lower level of biotic organisation as components - this is a feature we shall return to later.

The environment comprises external or internal circumstances that are causally relevant to action. Circumstances can be causally relevant to action in two ways: they supply more or less favourable opportunities for action, and they do so on an occasion when the circumstances obtaining on that occasion belong to the range of an ability in the agent's repertoire; on that occasion the agent has the opportunity to do something he can do. And, secondly, circumstances partially determine the outcome of an action. Consider the following structure familiar from decision theory.

In relation to inside temperature the outside temperature is

	higher	lower
open the window	inside temperature rises	inside temperature sinks
do not open the window	inside temperature remains constant	inside temperature remains constant

What happens if the agent chooses the alternative of opening the window? It depends on alternative circumstances or 'states of nature', the inside temperature will rise or sink depending on what the circumstances happen to be. But, as the table shows, the agent's choice of action is also important. Here is another example of the same kind.

	In relation to genetic disorders the child will be	
	affected	not affected
reproduce	birth of an affected child	birth of a child not affected
do not reproduce	birth of no child affected	birth of no child not affected

Similar examples pertaining to the internal choice environment will be given below.

The third determinant, goals, has a very rich structure. As regards this, a few comments will have to suffice here. The life plan is the goal which the agent has for herself, for what she should become or continue to be, a professional gardener, say, or a good parent, or a successful business man. Since these do not exclude each other, we have to allow for goals comprising projects that are ranked in order of importance by the agent. Given this, some projects are central and others less central in the agent's life plan profile. Normally each project is further structured into sub-projects which may be ordered in various ways. One such ordering relation obtains between two sub-projects when one cannot be finished unless the other is finished.

The three determinants form a system of adaptedness when they are evaluated from the point of view of each other. To illustrate this we look at two possible evaluation situations (for the value-theoretic concepts used in this section, see Österman, 1995). In the first, goals form the point of view for evaluating the repertoire and the environment taken together or severally. As such, goals constitute a source of requirements on the repertoire and the environment in the sense that the realisation of the goals requires the repertoire and the environment to meet certain conditions which may or may not obtain. The evaluation concerns the extent to which the repertoire and the environment meet the requirements. The value order used might be partitioned into the scale shown below.

better				
excellent	good	acceptable	bad	unacceptable
worse				

If so, evaluating the repertoire from the point of view of the goals is a matter of placing it in one of the five cells on the basis of the extent to which its exercise meets the requirements. The task is essentially the same in evaluating the environment from the point of view of the goals; here, too, the realisation of the goals requires favourable causal relevance in the environment, and the question is to what extent this is present in the environment. Since the range of an ability comprises conditions that may or may not obtain in the environment, it is quite natural to regard the repertoire and the environment as jointly forming a space of possibilities. Evaluating the repertoire and the environment together from the point of view of the goals is then a matter of judging to what extent these possibilities meet the requirements imposed by the goals. One way of expressing this kind of value judgement is to say that the agent is adapted to his environment with respect to his goals to the extent that his repertoire is adequate for realising the goals in the environment. The defining relation is the evaluative relation of adequacy, so instead of saying that the three determinants form a system of adaptedness when they are evaluated from the point of view of each other we could say that they form a system of adequacy.

The evaluation situation just reviewed may be reversed in the sense that it is possible to evaluate the goals from the point of view of the repertoire or the environment. As already mentioned, the repertoire and the environment determine a space of possibilities for action, and the question to be decided in the kind of evaluation now under consideration is how the goals are related to those possibilities. A goal may fail to respect the limitations to which the repertoire and the environment are subject. We then say that the goal is unrealistic. Alternatively, a goal may be evaluated on the basis of the extent to which it has the character of a challenge.

The state of adaptedness, understood as the system of adaptedness at a time or during a period of time, may in turn be evaluated. This evaluation is mediated by an evaluation of the actions determined by the state of adaptedness, and the point of view may be, for example, the well-being that attends these actions, or the happiness they confer or the meaningfulness of the life they compose. These points of view may be the agent's own or they may be external to the agent. An external point of view that frequently occurs is that of utility to others.

The state of adaptedness may be qualified by imposing restrictions on the terms of the relation of adequacy. For example, goals can be restricted to the central projects of the life plan, in which case the agent's adaptedness to her environment turns on her repertoire to realise the central projects in her life plan in that environment. Or the environment may be qualified, and so on.

The stage is now set for further developments. Given the setting - an action-oriented theory of human nature - health may be characterised as the repertoire regarded and evaluated as a component of the system of adaptedness. The evaluative character of health so understood is internal to the system of adaptedness and independent of points of view that may be used to evaluate the system itself. So, for example, there is no conceptual connection between health and happiness - a person may have good health during a period without being happy in his life as a whole during that period. However, a connection of a sort may enter by accident, namely when the person happens to have the pursuit of happiness as a central goal in his life plan.

There is no reference to disease in this characterisation of health. The field is therefore open for defining disease in terms of internal states or processes which tend to undermine health by reducing the capacity of the functional ingredients of abilities in the repertoire. The caution involved in talk about tendencies here seems to be justified, for even if it is assumed that every disease results in impairments of functions, it does not necessarily follow that disease undermines health. This is so because health is not simply the repertoire, but the repertoire regarded and evaluated as a component of adaptedness, and this perspective bring relations to the goals and the environment into view. It is the reduction of abilities to attain central goals in one's environment that undermines health, not the reduction of abilities as such.

A person's health position at the end of an interval of time may be judged to be better, or worse, or the same as his health position at the beginning of the interval. So there are three possible cases: improvements, deteriorations, and status quo. These are obviously relevant to the understanding of health care. B's action in relation to A may be said to be of type health care if and only if it affects A's repertoire and the effects are evaluated from the point of view of improving A's health, or avoiding a deterioration, or preserving a position which is (at least) acceptable. When health care is defined in this way it seems to be an extremely rich category.

Environment care and goal care are readily understood analogously. Care of adaptedness may be understood as the sum of these three forms of care. Environment care may be subdivided into care of the external environment and care of the internal environment. An important section of the latter is emotion care.

If medical care is understood classically, it is action intended to facilitate treatment and prevention of disease. Seen in this way, it may clearly be regarded as a part of health (repertoire) care. On the other hand, it does not exhaust health care for owing to the fact that abilities have ingredients other than functions impaired by diseases, there are non-medical forms of repertoire care, for example in rehabilitation work. Palliative care of the medical sort, for example control of pain, belongs to internal environment care.

How should the three goals specified by Nuffield Council be placed on the preliminary map of care just drawn? Goal one is a health care goal and in principle there is scope, and hopefully a great deal of scope, for medical intervention in this area using methods of gene therapy and other measures. It has often been observed, however, that the development of medical genetics involves a shift from therapy to predictive medicine because with increasing rate it is becoming possible to diagnose genetic disorders for which there is no known treatment. For the time being, and possibly for some considerable time to come, the contribution from medical genetics to the realisation of goal one is therefore restricted to a large extent to whatever contributions it can offer through prediction and prevention. As regards the latter, fear is often expressed that many societies will go for selective abortion as the method to be favoured.

We turn next to goal two, and we begin with an object of comparison. The agent knows how to use the telephone - this is an ability in his repertoire. But he does not have access to the number of the person he wishes to call. As long as this lack of information prevails the agent is deprived of the opportunity to call the person concerned. And giving him the number is a case of internal environment care. Allowing carriers of a given abnormal gene to make informed choices regarding reproduction should be understood in the same way. That is to say, it does not improve their reproductive capacity, but as an intervention in their internal environment it gives them an opportunity to use the reproductive ability on the basis of information about possible outcomes. Giving them this information is again a case of internal environment care, the value of which has to be judged by taking a number of factors into account, not least the attitude of the carriers.

There is an alternative way of looking at this matter. In the case used as an object of comparison the ability involved may be specified, much more narrowly, as the ability to phone a designated number. Giving the person the number he lacks would then amount to a case of repertoire care. Similarly, the ability to make an informed choice regarding reproduction may be understood narrowly in those terms. If so, giving a person information about an abnormal gene he is carrying would be a case of care of decision repertoire.

However, whether goal two is seen in the first or in the second way, it seems to have little to do with the practice of medicine. Even if we assume that the information to be conveyed has a medical origin, it does not follow that the use of it, in environment or repertoire care, is medical by nature.

Goal three is without doubt a goal for emotion care, which forms a part of internal environment care. It is well-known that emotions are extremely sensitive to information. To illustrate, assume that there are two states of the internal environment: i. that a woman wants to become a mother but not the mother of an affected child, and ii. that she wants to become a mother and is prepared to accept an affected child (cf. the decision matrices on p. 5). If she forms the belief that the child may be affected, her emotional response in case i. is fear of having an affected child; in case ii. it may be hope that she will be able to cope. If, on the other hand, she forms the belief that the child will not be affected, her response in case i. will be confidence and relief and in case ii. the same but possibly also increased self-respect. If instead of forming the belief that the child may be affected she forms the belief that it may or may not be affected, her emotional response will be anxiety, and so on. Emotion care is possible partly because of the cognitive component of emotions which admits the question "Why do you believe that ...?", and partly because why-questions are meaningful also when applied to desires which are components of emotions as well.

However, fear of a genetic disease, or anxiety, is not itself a disease any more than fear of becoming unemployed is a disease. These are reasonable responses to contingencies in life. Alleviating such emotions is not a medical task to be discharged by drug therapy. Suggesting otherwise is to engage in the rhetoric of medicalisation. The same remark applies to goal two, and to goal one if this is based on the notion that all health care is medical care, which is likely to be the case if the conception of health as the absence of disease is assumed.

In order to place the above remarks concerning emotion care in the right perspective two qualifications have to be made. First, emotional responses may be more or less appropriate, in kind or in degree. Fear or anxiety do not always call for alleviation, for there are circumstances in which such emotional responses are (most) appropriate. In such cases there may still be a place for care intervention, for example to confirm persons in their emotional responses.

Second, emotion care is not an end in itself. Its value is determined on the basis of the contribution it makes to the formation and exercise of practical judgement, i.e. the ability to identify, choose and implement desirable alternatives. This contribution takes the form of counter-acting the tendency of inappropriate or inordinate fear or anxiety to blur the perception of desirable courses of action and to undermine firmness and self-control in the pursuit of them.

Goal two should be assessed in the same way, i.e. from the point of view of the role its realisation plays in the formation and exercise of practical judgement. It is often assumed, usually tacitly, that the ability to make informed choices regarding reproduction is a necessary ingredient of the ability to distinguish between desirable and undesirable alternatives. The assumption is by no means self-evident. A suitable object of comparison would be the assumption that knowing the time of one's death is essential to the formation and realisation of a desirable life plan.

Properly understood goals one to three are attractive, but they tend to lose their attraction, or some of it, when they are used to support medicalisation, spurred on by geneticization.

REFERENCES

Boorse, C., 1977, 'Health as a theoretical concept', *Philosophy of Science,* 44, pp. 542-573.

Gelehrter, T.D. and Collins, F.S., 1990. *Medical Genetics,* Williams and Wilkins, Baltimore.

Lippman, A., 1992. 'Led (astray) by genetic maps: the cartography of the human genome and health care', *Soc. Sci. Med.,* 35, pp. 1469-1476.

Morriss, P., 1987. *Power. A philosophical analysis,* Manchester University Press, Manchester.

Nordenfelt, L., 1987. *On the Nature of Health. An Action-Theoretic Approach,* D. Reidel Publishing Company, Dordrecht.

Nuffield Council on Bioethics, 1993. *Genetic Screening: Ethical Issues,* Nuffield Council on Bioethics, London.

Österman, B., 1995. *Value and Requirements,* Avebury, Aldershot.

Pörn, I., 1984. 'An equilibrium model of health', in L. Nordenfelt and B.I.B. Lindahl (eds), *Health, Disease, and Causal Explanation,* D. Reidel Publishing Company, Dordrecht, pp. 3-9.

Pörn, I., 1993. 'Health and adaptedness', *Theoretical Medicine,* 14, pp. 295-303.

Whitbeck, C., 1981. 'A theory of health', in Caplan, A.L. *et al.* (eds), *Concepts of Health and Disease: Interdisciplinary Perspective,* Addison-Wesley Publishing Company, Reading, Massachusetts, pp. 611-626.

Wulff, H.R. *et al.,* 1990. *Philosophy of Medicine,* (2nd ed), Blackwell Scientific Publishers, Oxford.

Chapter 16

Genetic testing, genetic screening and privacy

JUDIT SANDOR
Budapest Political Science Department
Central European University
Budapest, Hungary

1. INTRODUCTION

For moral philosophers and lawyers one of the most embarrassing characteristics of genetics is that it touches intergenerational problems. It follows from the intergenerational nature of genetics that it is difficult to draw the line between medical and non-medical information produced by genetic testing or screening. Since the concept of medical confidentiality, and parallel to this privacy protection, has grown out of the elementary human wish to maintain control over the individual's own personal information, including medical data, it became a challenging task for philosophers and lawyers to define how far genetic data can be classified as personal information and how far it can be classified as information which affects spouses, children, the fetus or the planned child. David Heyd (1992) elaborates this scepticism when he claims that prior to any moral and legal discourse it is necessary to decide whether questions provoked by genetics are parts of ethics, or whether they belong to a non-moral sphere of deliberation and evaluation. It is not my purpose to challenge this position but it may be evident from my enterprise that genetic screening, testing and moreover fundamental problems of genetics are moral problems and those who choose to undergo these examinations are subjects of the general legal

R. Chadwick et al. (eds.), The Ethics of Genetic Screening, 181–190.

protection of persons. In contrast to ordinary medical diagnosis both genetic screening and genetic testing touch upon the questions of future generations and as a consequence they go beyond the sphere of moral and personal rights of one single individual concerned (e.g. counsellee). However, despite the special characteristics of 'genetic diagnosis' it does not follow that these methods are not subjects of moral and legal concern. Parallel to this, privacy as a legal concept deals with a special element of rights attached to persons: individuals who have legal capacity and whose personal rights are at stake. That close link between the right and the individual explains why jurisprudence has never dealt with the challenge to provide privacy protection to non-existing future generations. Genetics, however, raises the questions of privacy rights of future generations, since decisions made by the ancestor to request testing produces and discloses such information which may be proved later to be essential to the next coming generation. Moreover genetic information affecting future children may be stored and registered before the individual was born or even conceived since the parents' genetic make-up does determine the child's 'genetic book'. In the future the doctor may have to inform the patient that genetic data concerning her ancestor is available. It seems evident that in this case the temptation to know becomes significantly bigger in contrast to the situations where the decision to request genetic testing still has not been made. Someone may even argue that it creates significant limit on the individuals' choices.

So far I have applied the term of privacy protection as an equivalent of data protection. However, both legal scholarship as well as the practice of courts all over the world do provide some legal protection for several other aspects of our life. Boundaries between private and public spheres have been changed through history and they are still different in different regions of our present world. That explains why there still does not exist an established scope of the private sphere with special legal protection. Parallel to emancipation and modernisation in the Western societies boundaries of private and public spheres have been again and again remodified.

One cannot find an answer either if one looks at the origins of the term privacy. Etymologically privacy derives from the Latin *privatus* meaning withdrawn from 'public life' or 'deprived of office' and stems from the verb 'private' meaning to deprive. In contrast the word 'public' stems from *pubes* referring to adult males. If one looks at the history of privacy protection it is obvious that in the beginning, when the first judicial decisions tried to shape its scope, privacy was understood as a very concrete sphere of the individual life (such as her home, her letters etc.)

The history of privacy protection quickly became a process of development from protection against literal physical intrusion towards the less visible forms of interference. In the American jurisprudence Brandeis' view of an expanded right to privacy opened the door wide for numerous judicial decisions based on privacy. He also turned out to be a good predictor of the future when in his dissenting opinion in Olmstead he already foresaw the consequences of future technical development:

> The progress of science in furnishing the government with the means of espionage is not likely to stop with wire tapping. Ways may some day be developed by which the government, without papers from secret drawers, can reproduce them in court, and by which it will be enabled to expose to a jury the most intimate occurrences of the home. Advances in the psychic and related sciences may bring means of exploring unexpressed beliefs, thoughts and emotions....Can it be that the Constitution affords no protection against such invasions of individual security? (Olmstead v. United States, 1928).

The possibilities which genetic testing and screening can provide us now, however, have gone even beyond these fears.

The appearance of genetic testing and screening provides a new challenge for legal protection of persons because of the very special nature and scope of these techniques.

1. About 3% of all pregnancies result in the birth of a child with a significant genetic disorder or disability (WHO, 1995).
2. Hereditary conditions are irreversible therefore the information about the test results influences the entire life of the individuals and his/ her family.
3. Hereditary conditions are strongly connected to one of the most important and intimate spheres of the individual, reproduction. Knowing a test or screening result may affect the individual's choice in reproduction and raise a burden of responsibility for the new generation.
4. A decision of one single individual to require a test may affect many others: family members, future spouses and children.
5. Disclosure of genetic diseases affects the most intimate sphere of life: reproductive decisions.
6. Finally, the disclosure of partial or entire medical and non-medical information provided by test results, as well as decisions followed by testing, affect women disproportionately more.

Although I am aware of the substantial differences between the two types of genetic examination in my paper I intend to cover both genetic testing and screening. Genetic screening is applied to large-scale populations with no known excess risk to individual persons. Diagnostic testing differs from screening in regard to the population served. Whereas screening applies to populations with unknown risks to individuals, diagnostic testing is offered to individuals and families who are at higher than average risk because of family history of genetic disorder, history of environmental exposure, advanced maternal age, positive results of a prior screening procedure, or clinical signs in the persons to be tested. It follows from the differences between the two types of examination that they raise partly different ethical questions. Screening, for example, is sometimes carried out without informed consent since it is often part of routine medical care. But protection of privacy in both cases raises the same problems and also despite the obvious differences between the two types of genetic examination the dividing line between the two sorts of examination often seems to be not so sharp. The simple fact is that sometimes screening or testing is coercive, in the sense that the omission of it threatens the patients in question with the cancellation of some medical or insurance benefits or simply by imposing moral responsibility on parents who choose not to undergo testing. The external influence which favours testing in itself questions the rigid frontiers between testing and screening. In our present world conceiving a potentially disabled child is considered as wrong, sometimes it is regarded even as a form of cruelty on the side of parents, and as a consequence of this attitude parents who belong to the endangered group may feel a moral duty to avoid bringing a disabled child into the world. It is beyond the purpose of my argument to challenge the assumed or even real demand by the society for parents to procreate healthy children. This expectation may follow from utilitarianism, consumerism or perfectionism. Since I do not search the reasons of such demands nor do I make an attempt to challenge this demand. Nevertheless I think we should take into account this external demand since it may alter or at least influence the individual's voluntary decision on whether she asks for testing, consent to screening or not.

If the external demand for testing and screening does exist we have to see that it also means that privacy in respect of genetic screening generally involves not only the right not to have others know but also the right not to know about oneself. Genetic diseases have special characteristics which justify a different ethical consideration than other non-genetic diseases. First of all these diseases are regarded as irreversible and more or less objective conditions. The threat of developing symptoms affects the individual's life, especially in the case of late onset diseases. The genetic anomalies may be inherited by the offspring, they may affect the carrier and his or her spouse's chances of having a healthy child. The right to Privacy is generally meant to be the control over someone's information about herself. Genetic testing and screening undermine this notion in two ways:

1. genetic testing or screening of one individual does affect family members and often relatives who do not even have regular contact. They might live in distant places, may have different religion, education and live in a different society;

2. genetic testing and screening does not really facilitate a right to decide, since after results of the testing or screening are disclosed - in the cases of adults - treatment for genetic disorders may not be available, while in the case of ante-natal testing the only 'treatment' is a genetic abortion. In the case of premarital testing the 'treatment' includes not to get married or not to have children if, for example, both parents are proved to be a carrier. It follows that the individual control over her personal information should be guaranteed during the entire medical consultation and examination.

1.1 Screening should be voluntary and should be preceded by informed consent.

Protection of personal privacy and its limits occur in several stages of genetic screening and testing. The moral questions start with the case when a planned child's parents carry a genetic disease. Does it automatically limit the parents' rights to privacy and require action from the doctor? If one of the parents carry a genetic disease does it limit the privacy of the affected spouse?

1.2 Newborn screening

The primary purpose of mandatory newborn screening is to benefit the newborn through early treatment. It follows that it makes no sense or even that it can be considered as unethical to provide screening if treatment is not available or in the case of the so called late onset diseases. Newborn testing and screening, however, do have a substantial effect on parents who may have never wanted to be tested. Information on the test and screening result is necessary in order to provide proper care for the already born child but on the other hand the information does affect the parents' decision to have further children or not. Doctors involved in testing and screening have to understand that parents cannot be left out from the procedures but also that their entire life may be affected by the genetic information.

2.   PREMARITAL OR ANTE-NATAL TESTING AND SCREENING

2.1   Disclosure of Test Results to Spouses and Partners

Under normal circumstances if a couple intends to have children, men and women share information with their partners in order that both be aware of potential harms to a future child. Ethical guidelines usually treat men and women equally in testing and screening. Here, however strong asymmetry can be observed between the situation of the man and the woman. If the woman does not know a relevant information about her male partner and she carries the pregnancy into term when later on it turns on the baby is seriously defective her nine months, physical and mental sufferings are wasted. Moreover this tragic experience may prevent her from the chance and capability of being able to carry a child to term again. After the birth of a defective child women usually stay at home with them since women still have a special position as caregivers for children and especially children with disabilities. The needs of a disabled child in most cases result in the end of all individual life plans and career goals of the mother. A disabled child requires not only more emotional and physical care but also more financial support. Since women usually have less favourable access to economic resources than men do it follows that a woman with a disabled child is in a multiply disadvantaged position. Both pre-natal genetic testing and screening often include intrusion into the mother's body involving also painful and risky interventions. On the other hand if a woman either because of religious belief or simply because she fears the test, refuses to undergo ante-natal testing of her babies she may be regarded as irresponsible and selfish. The external expectation for perfect or at least healthy babies strongly questions the voluntary nature of the consent to ante-natal examination. The pressure for testing and screening may even increase in the future as genetic technology advances. However, one has to note that even if the test itself will not be invasive, still treatment of the fetus as well as the ultimate decision to terminate pregnancy will still jeopardise the woman's bodily integrity. On the other hand this will naturally solve the problems derived from the invasive nature of pregnant women testing.

However, we are not yet at this stage. The most common methods of obtaining genetic information on the fetus are amniocentesis and chorionic villus sampling, both of them invasive techniques with the risk of provoking spontaneous abortion. Amniocentesis is regarded as cheaper and as a consequence is more frequently used in Hungary. Amniocentesis is usually carried out during the 16-20 weeks' gestation. On the other hand chorionic villus sampling which can be carried out before the 12 weeks gestation involves a higher risk of miscarriage. Since we know the link between the mother's age and the elevated risk of delivering a child with Down's syndrome this finding has created the notion of so called 'aged pregnant woman' which in some countries means that the pregnant woman is above 30 or 35. There is no room here to explain the stigmatisation created by such a categorisation. It is however obvious that an external demand exists towards women to plan their life around the genetic risk. If in such a short period of time 'aged pregnant women' already constitute a threatened group within the group of all pregnant women a further question may be raised: whether the other findings of genetic tests will place the same or even more grievous burdens on parents who fall into a disadvantageous genetic group. Can this process lead to the time when natural pregnancy will be a privilege for some families whereas for several endangered groups decisions on pregnancy or continuation of pregnancy will depend on genetic examination? Here another reasonable fear can be mentioned: whether the cost-benefit calculus will be assessed prior to birth or conception. It is common sense that the health care of handicapped or sick children requires more health resources, also we know some genetically inherited diseases which require later on substantial financial resources. But since there was until recently no possibility to detect the potential patients until after birth or even only when the disease manifested itself, no one considered it morally acceptable to raise the question whether the costs spent on this group of patients should be avoided or eliminated. However once a health care system is able to 'screen out' the potential 'big health care consumers' due to their parents' genetic testing and screening it will be difficult to avoid the temptation to promote testing which in every case includes an implicit evaluation of whether the pregnancy is desirable or not for the society. Unfortunately for those who plan allocation of health resources purely on utilitarian grounds this prior birth or prior conception assessment makes sense and it follows that under these health regimes the economic pressure for testing and screening of certain groups may grow by finding more genetically linked expensive diseases. If testing or screening together with the costs of false negative and positive results will cost less for a health care system than providing significant health resources for decades for large numbers of children and adults suffering from genetically detectable diseases then the threat will become real. This basically eugenic thinking is supported by the growing judicial appreciation of quality of life considerations introduced by so called wrongful life cases where children sue their parents and the doctors on the ground that their life is miserable and their birth should have been avoided. I do not want to enter into the analysis of the absurdity of the children's claim in questioning their own existence, I simply want to highlight that 'quality of life' measures already have practical, legal implications which show not only in the direction of providing guarantees for the individual to have access to a certain level of care but also the right to question health services on the ground that even with maximum efforts they produce miserable lives.

The findings of the tests or screening again place the burden on women and the difficult decision whether to undergo abortion (sometimes with Caesarean section due to the advanced pregnancy) or to continue to carry the disabled child to term, perhaps without family support and expecting the delivery of a disabled child. If one takes into consideration the potential false positive or false negative result of the test the emotional burden is close to unbearable. On the other hand if a woman refuses testing, she might face lifelong self-blame and isolation in the family.

While high-tech diagnostic medical technology has developed around the pregnancy the basic features of pregnancy remained unchanged: it lasts for about 9 months, it is a very important period in the family life, in some societies successful pregnancy is the only success indicator for women. On the other hand even when it is desired it involves emotional and physical suffering, delivery of the baby either in a surgical or in a natural way is painful and involves the risk of maternal or fetal death or irreversible or reversible illness. Also if for whatever reason a pregnancy is unwanted either based on the social and emotional conditions of the mother or so called fetal indications, in order to terminate it, it is inevitable to conduct a medical intervention with intrusion into the mother's body. Termination of pregnancy is also termination of fetal life and as such it is never an ordinary operation but it is always a mentally and emotionally demanding procedure. When any screening or testing in the pre-natal stage occurs one should bear in mind that the advanced techniques in the field of testing and screening have not at all ameliorated the pregnant woman's position - she faces the same destiny just like centuries ago - she worries about the health status of the embryo and her own physical safety. What really has changed in this respect is the earlier notification of the child's health condition. With the significant drop in the number of children in Western societies pregnancy more and more becomes a success oriented project where the small number of children has to be produced perfectly and in due time. Emily Martin analyses the production metaphor in relation to pregnancy (1987). She also places the doctor in the model as a supervisor; the woman might be a labourer whose machine produces the 'product' babies. As a consequence of the existing social demand to decrease the ratio of disabled children it is difficult to provide suitable and applicable guarantees to ensure free and voluntary consent to genetic testing and screening. The right to self-determination is strongly linked to the concept of human dignity. But no one can exercise an autonomous decision without being previously informed. There is a golden rule in medical ethics: the more complicated and the longer effect a medical intervention has the more considerate and detailed information should be provided for the patient. Genetic testing and genetic screening in this sense requires additional legal guarantees since disclosure of the findings of medical examination after the examination in an inevitable consequence of the examination, it follows that a prior assessment of risks has to be completed before a decision is made on the examination itself.

The effects of 'genethics' will alter our notion of the private sphere and also the means of legal protection of privacy rights. One of the obvious and already wide-spread mechanisms is data protection which imposes a control over how information is obtained, used, processed and disclosed. However it does not provide sufficient guarantees for the protection of genetic data, since once genetic data has been obtained, the notification about the potential danger for children and relatives or disclosure for the partner is often regarded as a moral duty. Moreover genetic screening and testing may reveal also secondary findings such as non-paternity. Also once a consent is given to one specific testing it is difficult to avoid the risk that some other undesired findings will be registered. Special attention has to be paid to the fact that despite some promising new techniques (such as gene therapy) at the moment medicine can offer only very little help for genetic diseases. It follows that the information on testing and screening results does not really authorise the individual to do anything and only in case of premarital testing provides a chance for the individual not to do anything, for example to avoid natural conception. Genetic testing and screening undoubtedly support the various forms of medically assisted reproduction around which a significant market has already developed.

Harms and benefits deriving from genetic testing and screening do affect third parties and the whole society in general. One of the often overlooked consequences of genetic testing and screening is that it implicitly includes a bad message for the relatives and friends especially when genetic disease is found. The message is often read by the patients having genetic diseases as the following: the society does take care about me does tolerate my genetic deviancy and has even provided some social and health benefits but if it could have avoided my birth it would have done it. I am aware that it is not necessarily the case with every form of genetic testing and screening, however, when after pre-natal testing abortion is recommended or just mentioned as an alternative this hidden message becomes evident. That unfavourable third party effect supports the position that until therapy becomes available testing and screening should be considered as an irregular treatment or even as research which requires special guarantees in respect of initiating testing and screening.

REFERENCES

Heyd, David, 1992. *Genethics Moral issues in the Creation of People,* University of
    California Press, Berkeley.
WHO Hereditary Diseases Programme, 1995. *Guidelines on Ethical Issues in Medical
    genetics and The Provision of Genetic Services,* Preface, WHO, Geneva.
Martin, Emily, 1987. *The Women in the Body, Medical Metaphors of Women's Bodies,* Birth
    Open University Press, Milton Keynes, pp. 55-67.
Olmstead v. United States, 1928.

Chapter 17

Reconciling liberty and the common good?
Genetic screening in the Republic of Ireland

DOLORES DOOLEY
Department of Philosophy
National University of Cork
Ireland

1.   CONSTITUTIONAL RIGHTS AND THE COMMON GOOD

The value of individual liberty in Ireland has been, at once, legally dominant and yet culturally divisive. If one were to judge from the many legal decisions given since the adoption of the Irish Constitution in 1937, respect for the liberty rights of individuals has been dominant. These judgements have specified a series of rights for individuals which are derived from, though not explicitly mentioned in, the Irish Constitution. In their judgements since 1965 in *Ryan v. Attorney General*, Justices have argued that an expansive view of the Irish Constitution is appropriate.[1] This means that the statement of explicit rights in the original Constitution cannot be understood as exhaustive and, that with judicial interpretation, other latent, unspecified rights can be enumerated. In article 40.3.1 personal citizen rights are given protection:

> The State guarantees in its laws to respect, and, as far as practicable, by its laws to defend and vindicate the personal rights of the citizens.[2]

In addition, Article 40.3.2 might be potentially relevant to the important question of the Constitutional protection of genetic material as property of individuals and privacy of genetic information. Article 40.3.2 states:

R. Chadwick et al. (eds.), The Ethics of Genetic Screening, 191–205.

The State shall, in particular, by its laws protect as best it may from unjust attack and, in the case of injustice done, vindicate the life, person, good name, and property rights of every citizen.

Since *Ryan v. Attorney General* in 1965, the Irish courts have identified at least eighteen unenumerated rights constitutionally protected by 40.3.1. Thus, apart from the significant implications of European Union law on the legal protection of rights in Ireland, the Constitution guarantees 'Fundamental Rights' to citizens in Articles 40 to 44. At the same time, the Irish Constitution borrows strongly on a concept of the 'common good' which occurs in articles of the Constitution as a normative criterion against which behaviours will be evaluated as politically and morally permitted. Article 6 makes clear:

All powers of government, legislative, executive and judicial, derive, under God, from the people, whose right it is to designate the rulers of the State and, in final appeal, to decide all questions of national policy, according to the requirements of the common good.

Article 43.2 further reinforces the concept of the 'common good' as requiring compatibility with ascribed liberty rights for individuals:

The State, accordingly, may as occasion requires delimit by law the exercise of the said rights with a view to reconciling their exercise with the exigencies of the common good.[3]

The case of *Ryan v. Attorney General* not only affirmed the legitimacy of interpreting unenumerated rights based on the Irish Constitution but stated more explicitly what theoretical framework would provide the basis for justifying the positive ascription of individual rights. The 'rights' that might be specified would have to 'result from the Christian and democratic nature of the State'.[4] In addition, Article 44.2.1 of the Constitution prescribes that certain behaviour and choices of citizens are constrained by reference to other norms: 'freedom of conscience and the free profession and practice of religion are subject to public order and morality'. So any discussion of individual rights to health services within the country are subject to being measured against 'public order and morality' which can, in principle, function as constraints on conceding certain personal rights to certain health care services such as genetic testing and/or counselling provisions both ante-natal and post-natal. Such rights may be measured against a vision of the 'Christian and democratic nature of the State', public order and morality and a highly contested concept of the 'common good'. For decades since Irish Independence in 1922, the terms, 'morality', 'public order', the 'Christian nature of the State' and the 'common good' were understood as related concepts, each mutually interdependent and ethically coherent.

During the last fifteen years in particular in Ireland, debate has grown about a number of medical-ethical issues including genetics services. These debates have provided impetus for State and citizen efforts to achieve clarification on health care rights for individuals; along with the rights deliberations are questions about possible revisions in the meanings of the 'common good' as a constitutionally specified constraint on rights of individuals. Such debate also raises issues of meaning surrounding such concepts as the 'life, person, good name, health and property rights of every citizen' and the extent to which these might appropriately apply to the genetic material and DNA property which is so essential to one's personal identity. A question that only promises to become more acute is whether the deliberations of the body politic and the Government of the State are going to concede the liberty right of individuals to have available to them information on *their own* genetic identity. The proviso would be that, to the extent that it is possible to control this, individuals *would choose to have this information.*   A resolution of this liberty-right's issue will require that the determination of the 'common good' be re-interpreted to encompass a more pluralist citizenship than was factually the case when the Irish Constitution was drafted in 1937. It is increasingly recognised that a political goal to aim towards is that of achieving a balance between, on the one hand, a defensible and pluralist interpretation of the 'common good' and individual rights which would provide liberty of choice to use or refrain from using a wide range of public health-care services not previously provided in the State. This project does not suggest two separate and distinct realities: 'the common good' and 'individual rights'. The effort of contemporary Irish citizens to negotiate a more pluralist conceptualisation of the 'common good' would precisely imply expanding the range of liberty rights to meet value differences of legitimate moral perspectives.

I have argued so far that the contested but correlated values of the 'common good' and 'individual rights' largely define the public debate about access to genetic information and the extent of health care provisions which will be made available to citizens of many religions and no religions at all in the Republic of Ireland. The question of costing such services in the State is indeed relevant but is not the primary focus of debate in providing such services.

A Roman Catholic tradition has, since Irish Independence in 1922, provided a strong cultural ideology for interpreting the common good. To the extent that the concept of the 'common good' is primarily interpreted by reference to the deontological ethics and moral teachings of the dominant religious Church, to that extent the question of respect for the diversity of ethical views of citizens much, more prominent today, becomes a central and contested moral and political issue.[5] The membership of Ireland in the European Union has effected changes in the cultural identity of many Irish citizens and alterations in the perception of the Irish Government as more than an Island government. The European Communities Act, 1972, as amended by the Single European Act 1992, provided for the integration of Irish law with law of the European Union. The effect of this is to provide that EU law has the paramount force and effect of constitutional provisions. EU law thus is also one of the determining features of the Irish medical law picture. Article 129 of the Maastricht Treaty provides that the Community shall:

> ...contribute toward ensuring a high level of human health protection by encouraging co-operation between Member States, and if necessary, lending support to their action. Community action shall be directed towards the prevention of diseases, in particular the major health scourges, including drug addiction, by promoting research into their causes and their transmission, *as well as health information and education.*[6]

The access of citizens of Ireland to the services of other member states in the European Union has been a major factor contributing to an increasing pluralism even against the cultural backdrop of cherished religious traditions.[7] This uneasy but potentially creative alliance between a dominant religious tradition and moral ethos with increasing value pluralism that clearly and publicly include non-religious perspectives within the culture is further teased out in what follows.

2.   VALUE PLURALISM AND THE COMMON GOOD

Voices of professionals and lay people alike are beginning to mobilise for voluntary access to information on one's health. Such access, especially to genetic testing where there is clear family history of inherited genetic diseases, should be more widely available with necessary counselling provided than is presently the case where such services are dominantly accessed by medical professionals. Citizens are pursuing the objective to **at least have some choice** in seeking to know or not know one's personal genetic identity and the genetic identity of potential offspring.

> It is not enough to have small groups of professional people making decisions on fundamental issues which will change our lives forever. What is done with this knowledge is for society to determine. Professionals have the power only because they hold the information. Society needs to become genetically literate, public debate needs to happen so

that people can make informed decisions about their lives and thus maintain control over and use of genetic information.[8]

Publicity is negligible about the existence and services of the National Genetics Unit or, indeed, about private genetic testing services in the State. There are reasons for this which have to do with the national memories of divisive protests and emotive rhetoric during a series of public debates about: the introduction of a comprehensive contraceptive service (1981), acrimonious debate over a Constitutional referendum to allow the unborn to be named a 'citizen' of the country (1983); the provision of information for abortion services outside the Irish State (1995), and the availability of sterilisation for women without requiring approval of an ethics committee. In the public domain where deliberative debating is attempted, many adaptive mechanisms are constructed to avoid the more unsettling or hurtful elements of public discussions. There is a tendency to keep certain provisions in health care services (relatively) discreet in the belief that publicity may well meet with resistance, disruption and discontinuation of services that have been put in place. Such services are desired by some and feared by others. The essential point is that public debate is not seen by many as an unqualified good especially from the interpretation of the common good which would, for normative reasons associated with a substantive 'common good', delimit individual decision-making by competent adults. The concept of democratic politics envisaged here involves:

> ...*public deliberation focused on the common good*, requires some form of *manifest equality* among citizens, and *shapes the identity and interests* of citizens in ways that contribute to the formation of a public concept of common good (Cohen, 1997, p. 144).

A 'public concept' of the common good is in creation. It is a slow process which will require careful defense of revisions to the extent that these changes are seen as diverging from the theologically informed 'common good' of the 1937 Constitution and subsequent State legislation. An example of dissonance between individual rights and competing versions of the 'common good' is evident in the continued public silence about services for pregnant women to have amniocenteses. Amniocenteses are not 'publicly acknowledged' to be available even in public hospitals of Ireland. The rationale for this invisibility of the service (known by insiders to exist in only several hospitals) is that, there is no point in providing amniocenteses unless the option of abortion is provided in the State. Abortion is unavailable in the country and legislation currently does not exist to provide for such. Therefore, no service that would provide information that would facilitate making such a choice of abortion should be provided. What is apparent is that the restriction on information is being defended on the grounds that such information could lead to decisions that might possibly collide with a traditionally preferred Christian [Roman Catholic] view of the 'common good'. Effectively, citizens of the State are not given a 'right to know', with accompanying service facilities to come to know, where such knowing might contravene a traditional set of deontological and religious values. This is a highly contested repudiation of the pluralism that is an inevitable and visible part of Irish life today. There is also a note of unrealism in the assumption that citizens are not procuring amniocenteses outside the State and this, on the explicit advice of their doctors. Some women are seeking amniocentesis services and making decisions about whether to abort or not. Research has not been done to determine the extent of such interest but anecdotal information would suggest the numbers of such women are increasing. But the denial of ante-natal testing services within the State on grounds of no abortion services seems to achieve a dubious moral high ground on these matters but at the expense of denying the citizens' value pluralism.

The justification for lack of amnioncentesis mentioned above also lacks credibility according to Senator Mary Henry, M.D., who points out that older women are making abortion decisions in ignorance and because of exaggerated expectations of having a handicapped child.

> Most of these women do not have amniocentesis first and abortions are carried out because of lack of information and discussion on the real risk of maternal age....if amniocentesis is available in Ireland at least counselling can take place with those who normally are the woman's medical advisors (Henry, 1996).

People opposed to genetic counselling often explain their position by claiming that such counselling and subsequent testing could be a prelude to abortion. However, what is overlooked is the fact that couples may simply be talking about the important value of having informed pregnancy choices. Yet there is reluctance that individual rights might be exercised beyond this benign end and information would lead to alleged 'discrimination' against newborns with Down's syndrome, cystic fibrosis or other inherited genetic diseases. The concern is not trivial that publicly acknowledged individual rights to genetic services may almost be collusion in possible choices for abortion where a range of disabilities is genetically indicated. This right to pursue selective abortion without qualification (eventually extending to gender selection) is seen by many as a position at variance with the good of the community in its efforts to prevent discrimination of any group of individuals. This latter position of expressed caution about the genetic slippery slope is a view shared by more than a minority of citizens. However, the fallacy in the argument is that, while alleged discriminatory decisions to abort are possible when certain genetically inherited diseases are indicated, by excluding genetic testing altogether, responsible reproductive planning [whether to conceive] is also often curtailed.

In brief: the argument for strong constraints on the availability of genetic information is a complex religious-political one illustrating a strong traditionalist and theological interpretation of the common good: where information might potentially be used to make choices not compatible with the teachings of a dominant religious position, facilities to provide that information should be discouraged. This interpretation is validated by extensive debate about a Constitutional Amendment passed to allow for the State to distribute information relating to services lawfully available in another State. In 1995, Abortion Information legislation was passed giving general practitioners and others (Well Women Clinics) the right of providing information on where abortions can be obtained outside the State.[9] However, general practitioners cannot make appointments for women to have a termination. Robust disputes about the provision of information on abortion services in other EU States led to some public concern that individuals do not have a right to information if use of that information, even outside the country, might lead to a decision to destroy foetal life. The status of foetal life was reinterpreted when, in 1983, a national referendum adopted a Constitutional Amendment, 40.3.3. which elevated the unborn to that of 'citizen' whose right to life is seen as *equal to* the life of the pregnant woman.[10] Again, a strong deontological and pro-natal interpretation of the common good could override alternative claims for women's right to ante-natal information and the right of women to make the final choice regarding continuation of pregnancy.

Public discussions about the incidence of major genetic abnormalities in the Republic have given visibility to the need for more lay education in genetics and more understanding of available private or public genetic services. For example, individuals affected by muscular dystrophy have voiced concerns in support groups saying that they ought to be allowed genetic testing precisely so that they have the power to make choices. Greatly constricted services are preventing that information from informing choices for affected families. The tensions between individual rights and the common good are complex. Should limits to the individual right to know one's genetic identity be decided on the basis of dominant religious views that may not be shared in their entirety by the citizens of the country? Should the 'common good' include a recognition of value pluralism and provision of public policies that would allow a respectful allowance for choice of life styles consistent with that pluralism?

3.   GENETIC MATERIAL: WHOSE PROPERTY?

Adjudicating these questions requires public/social debate about the philosophy and functions of the genetics services in the country, probing inquiries about how the genetics services should be understood and how certain parameters on these services should be determined. Ironically, there is provided in the Data Protection Act a statement favouring a 'right to know' of all personal data held about one's self in any state centre or institution.[11] Corresponding to this is a protection of confidentiality and privacy concerning the institutional holding of data (including genetic or DNA data) about individuals. This position is in keeping with the recently approved Convention on Human Rights and Biomedicine. (1997)[12] Yet, not surprisingly, individual citizens in Ireland have only 'qualified control' of information about their person. Exceptions are mainly two. The first is uncontroversial and refers to disclosure of data to third parties where third party harm (general society or identifiable third party) is judged to be a substantiated probability if such data is not communicated. Efforts should normally be made to achieve consent from the individual for disclosure. However, there is another exception to requirement of consent for disclosure of confidential materials which is more controversial. This waiver of consent concerns medical data on an individual (including genetic data) which would be used in research but under conditions where the research would not reveal the identity of the patient. This waiver on participants' consent would normally have to be checked through an authorised ethics committee or Institutional Review Board assessing research programmes.[13] This exception on consent assumes endorsement of the belief that the larger good of 'human well being' served by scientific research is a value which takes priority over individual consent to use of their genetic materials. The logic also implies that is that one's 'genetic material, one's genetic property' can be utilised without having the possibility of gaining information on research results that could be potentially significant for decision-making. Where research is conducted using DNA samples from hospital patients, there may be no mechanism for tracing individuals who have provided the samples. Nevertheless, some of these individuals will be found to have a genetic disposition to certain diseases, the occurrence of which in themselves or their offspring might be ameliorated or otherwise prevented if they knew and could seek proper medical help. Should research which uses individuals' DNA materials not have some communication links established with individual data contributors? These questions which require adjudications between individual rights and interpretations of the common good should be seen as important elements in the social changes in Irish society and its active participation in the European Union. A minimisation of public debate, however complex the motivations, works to give power to a few and deprives many of services that could be very important to health care and reproductive decisions. Minimised public debate can work to constrain public demands for even relatively non-controversial developments in genetic services.

4.   GENETIC TESTING IN IVF EMBRYOS?

The Irish Institute of Obstetrics and Gynaecology is one body responsible for policy development on in vitro fertilisation and possibilities for genetic screening within that process. Revised guidelines on IVF of April 1992, state explicitly that 'under no circumstances should IVF be used to produce or store human embryos for research purposes' (Medical Council, 1994, p. 63). The Institute continues to debate the question of whether to allow freezing of fertilised embryos not solely for research purposes but for simplification of procedures of hormone treatment and ova retrieval to allow for future implantations of the couple concerned. There has been no change in the standing decision to disallow embryo freezing. This decision shows consistency with the underlying philosophy of not risking the destruction of human embryos if genetic tests give 'undesirable results' or if a couple chooses not to have further children and policy decisions might state that unused embryos should be destroyed. Here the competing interpretations of the common good show dissonance. Those who would look for expansion of IVF provision argue that the protection of embryonic human life is not absolute but can be qualified with respect to other entitlements to choose whether one will have an 'affected' child. This position would argue that it is better to make these decisions at this early multi-cellular stage of embryonic life rather than later in the gestation period when one might choose an abortion following amniocentesis. A strong pronatalist philosophy is expressed in the most recent encyclical of Pope John Paul II *Evangelium Vitae* which gives voice to a deontological religious view on pre-natal diagnostic testing which is endorsed by a non-pluralist view of the common good: When these

techniques are used with a eugenic intention which accepts selective abortion in order to prevent the birth of children affected by various types of anomalies...such an attitude is utterly reprehensible, since it presumes to measure the value of a human life only within the parameters of 'normality' and physical well-being, thus opening the way to legitimising infanticide and euthanasia as well.[14]

These issues serve to illuminate the diverse interpretation of liberty to make judgements and decisions about life, its early preservation and related judgements about 'quality of life' in efforts to make reproductive decisions based on genetic testing. These morally weighted decisions are very relevant to the prognosis for developments in genetic screening even at a minimal level.

5.   MANDATORY SCREENING FOR HEPATITIS B

A new compulsory screening program for medical students is expected to be introduced in 1997. Because of controversial elements in the screening proposals, they are likely to generate wider debate about genetic screening. The proposal, expected to come into effect for admissions to Medical Schools in the Republic in autumn 1997, will be mandatory testing for hepatitis B for all prospective medical students. Ireland's five medical and dental schools have taken this decision explaining that this decision was the result of "their moral and ethical responsibilities to protect both their students and patients". The Irish Medical Organisation, representing the majority of doctors in the country claims it has "no overriding objection" to the introduction of testing but it is concerned about whether students will have to pay for it! In addition the Irish Medical Organisation said it would try to ensure that the issue was explained 'sensitively but realistically to students'. At present, an advisory group established by the Minister for Health is still in consultation to consider the issue of whether, in addition, to mandate testing of practising doctors. To date this has not been recommended. The president of the Irish Medical Organisation made their concerns explicit: 'What we are concerned with is public interest and public health' (*Irish Times*, 1995). But if this is the case then perhaps another look at testing of registered doctors would be required.

Given the well established Irish constitutional right to bodily integrity, student representatives questioned the constitutional validity of mandatory testing. The students' appeal to the Constitution is an appeal justified by historical precedent.[15] The right to bodily integrity could be envisaged as providing grounds both for a legal appeal *against* mandatory testing but also *in favour* of State services for individuals to genetic counselling and testing. How interpretations of the 'common good' would fare in these potential appeals would remain open to further political debate.

A second concern voiced by mandatory Hepatitis B testing is that this compulsory testing might be the thin edge of the wedge - HIV testing or wider genetic testing by the back door. People going for testing who are part of high-risk groups (homosexuals) could be subject to subsequent discrimination. They could be asked questions not relevant to the testing process. The well grounded assumption is that Colleges will be keeping records. The issue of confidentiality and risks to student integrity cannot be underestimated. The earlier discussion of provisions in the Data Protection Act would provide some assurance of a prima facie confidentiality of test data. However, well wrought interpretations of the common good could, conceivably, result in the over-riding of personal rights to privacy.

6.   GENETIC CONFIDENTIALITY AND INSURANCE

The policy making body for medical registration, The Medical Council takes seriously the need for patient consent to any distribution of information to third parties (not likely to be harmed). This explicitly refers to insurance companies, employers and, one could add without changing intent, educational institutions. One of Ireland's senators, Jim McDaid, spokesperson for Equality and Law Reform (1997), warned recently that 'genetic tests for life assurance might create an uninsurable underclass'.[16] McDaid pointed out that clients with diabetes already pay double or treble the usual rates for life cover. McDaid's position is not unique but represents an increasing worry about the possibility of insurance companies introducing mandatory genetic testing especially for life coverage or mortgage protection. What may be sought as national legislation is the creation of community rating for a genetic pool. In other words, what is sought is a law that will require uniform premiums for life cover, regardless of a customer's genetic make-up. McDaid argues that 'far from being a *Star Trek* issue, genetic rating is a reality.'

Professor McConnell from the Department of Genetics at Trinity College has made a number of public interventions about genetic technology including the query about insurance. He is most adamant that emotional treatment of genetics in the media can be more damaging than anyone realises in obstructing the positive developments in services that can prevent serious, life-long illnesses. McConnell claims that we close our eyes to the developments in genetics only by incurring serious risks to future health and well-being.

> We are fast approaching a time when the aspects of our lives which we used to leave to
> 'fate' will be controllable in a laboratory. Yet our instinctive fear of these issues is making
> many of us in the Republic close our eyes to these developments (Holmquist, 1994).

McConnell is not sanguine about compulsory requirements for genetic testing from insurance companies. The time is not far off, he claims, when the genetic tendency for heart disease or certain types of cancer will be diagnosed in childhood, a development which he thinks should be welcomed since it will enable people to practise preventive medicine. However, few deny that this is a two-edged sword which the public needs to be aware of. In the absence of government controls, insurance companies could insist on blood tests which would reveal applicants' genetic risks of cancer, heart disease and other illnesses and, as a consequence, 'load' their premiums accordingly. This would put life insurance out of the reach of many and the contingencies of socio-economics and class differences would cause some to suffer more than others from such insurance requirements. Equally foreboding is the prospect that employers could hire and fire based on such blood tests, arguing that people at risk of certain conditions were not suitable for certain types of work. These possible eventualities for the use of genetic testing are not arguments for failure to develop genetic services. This would be counterproductive. The detailing of possible threats to individuals' privacy of genetic information is, rather, an argument for more directed and explicit government controls, more explicit than the existing Data Protection Act. Such government legislation needs to be put in place even at this time, in advance of sufficient public education in genetics and even with the country's provision of very limited access to genetic services.

7.   MANDATING DEBATE: MEMBERSHIP IN THE COUNCIL OF EUROPE

Since 1948 Ireland has been an active member of the Council of Europe and, for over 40 years, has been a signatory to the European Convention on Human Rights. Members in philosophy, medical ethics and law departments have been studying the Draft Convention for the protection of Human Rights and dignity of the Human being with regard to the application of biology and medicine (Strasbourg, July 1994). What was prominent in this Draft Convention and has been retained in the final form are the urgent and repeated recommendations for equitable access to health care but also "appropriate public discussion of the issues related to medicine, science and ethics".[17]

Regarding choices about the institution of genetic testing/screening the Draft Convention urges:

> Whenever choices are involved in regard to the application of certain developments, the latter must be recognised and endorsed by the community. This is why public debate is so important and is given a place in the Convention (Draft Convention, 1994, p. 13).

Member States in Council of Europe are asked several times in this document to

> create greater public awareness of the fundamental questions raised by the application of biology and medicine. Society's views must be ascertained as far as possible with regard to problems concerning its members as a whole. To this end, appropriate public discussion and consultation are recommended. The word 'appropriate' leaves the Parties free to select the most suitable procedures (Draft Convention, 1994, p. 31).

The debates about genetic technology in Ireland have evolved slowly but are helping to adjudicate a set of agreed basic values harmonising community concerns with individual liberty rights. This debate is increasingly seen by many as unavoidable even anticipating some challenges and possible rancour that abuses of genetic technology would outweigh the benefits. Such debate would inform the process of developing genetic technology. Public debates, often of a medical-ethical nature, are testing public receptivity to different interpretations of the public good, relative weightings of liberty, equality of persons, and the role of cultural-religious content in shaping important notions such as 'quality of life', 'the good society' and 'public morality'. Public invisibility of genetic services and closure on open debate need to be constructively surmounted if the indisputable reality of value pluralism is to be given recognition. Tolerant respect for such pluralism may be the essential ingredient of an interpretation of the common good capable of being reconciled with individual liberty rights and capable of guiding health care provisions into the year 2000.

NOTES

1 Ryan v. Attorney General , 294 Irish Reports, 294. (1965) The plaintiff in this case claimed that the flouridation of her water supply was harmful and interfered with her right to bodily integrity, a right which is not specifically mentioned in the Constitution but the Justice claimed was among the natural rights of a person.

2 Bunreacht na hÉireann (Constitution of Ireland) enacted by the People, 1 July, 1937.
3 See Clarke, Desmond. 1984, Church & State, Cork University Press, Cork, especially chapter 6, 'Human Rights and the Common Good'.
4 Ryan v. Attorney General, [1965], Irish Reports [IR], 294 at p. 312.
5 See Dolores Dooley, 'Expanding an Island Ethic' in Lee, J. J. (ed), 1985, Ireland: Towards a Sense of Place, Cork University Press, Cork, Ireland, pp. 47-65. The argument given here is that, with increasing pluralism and multiculturalism as part of the living context of Ireland, the dominant Roman Catholic Church can win credibility and leadership providing they present positions as linked to fundamental human values rather than biblical interpretations linked historically with the RC position. The challenge is to take a pro-active approach to engage in public discussion to make visible the fundamental values of traditional moral positions; if this is taken as a project, it may transpire that these basic human values might would be shared by many outside this one tradition.
6 This text is cited in Appendix E of Tomkin, David and Hanafin, Patrick, 1995, Irish Medical Law, Round Hall Press, Dublin, p. 287. The italics are my own.
7 See Anderson, Elizabeth, 1993. Value in Ethics and Economics, Harvard University Press, Cambridge, Mass.. Anderson's study provides excellent theoretical background on a pluralist theory of value which is at issue in this discussion. Arguments are given to show that adjudications of boundaries of autonomy, liberty and social constraints on these goods are part of political negotiations and determinations of what makes life worthwhile.
8 McCall, Thérèse, 1992. 'Our Fate is in Our Genes', The Irish Times, 7 January. McCall is a lecturer in Biochemistry, University College, Dublin.
9 The Regulation of Information (Services Outside the State for Termination of Pregnancies) Bill, 1995. This Bill was also referred to the Supreme Court by President Mary Robinson to insure that the provisions of the Bill were not repugnant to the Constitution and especially article 40.3.3. guaranteeing equal right to life of the unborn and pregnant women. The Supreme Court judged the Bill safe. See also the booklet prepared by the Irish College of General Practitioners, Training Programme & Information for General Practitioners (In Response to the Termination of Pregnancy Information Act 1995), adopted on 14 May, 1995.
10 Bunreacht na hÉireann (Constitution of Ireland), op.cit.
11 See Ireland's Data Protection Act, 1988. Government Publications Office, Dublin.
12 Council of Europe Treaty: Convention on Human Rights and Biomedicine, Oviedo, 4 April, 1997. This convention is a reduced version of the earlier Draft Convention which I find more explicitly committed to recommending public debate in member countries on matters of human rights and biomedicine.
13 Secretariat, Clinical Sciences Dept. Cork University Hospital, Wilton, Cork Clinical Research Ethics Committee of the Cork University Teaching Hospitals: Policies and Procedures Manual.
14 Pope John Paul II, Vatican Sacred Congregation for the Faith, Evangelium Vitae, p. 114.
15 On the right to bodily integrity as a constitutional unenumerated right, See Ryan vs. Attorney General (1965), op.cit.
16 See column reported by Gail Seekamp about genetic test warnings in The Sunday Business Post, 23 February, 1997.
17 Draft Convention, 1994, Strasbourg, Article 23. The final formulation of the Convention on Human Rights and Biomedicine of April 1997 is cited above.

REFERENCES

Cohen, J., 1997. 'Deliberation and Democratic Legitimacy' in Robert E. Goodin and Philip Pettit (eds), *Contemporary Political Philosophy,* Blackwell, Oxford

Draft Convention on Human Rights and Biomedicine 1994 Strasbourg: Council of Europe

The Irish Times, 1995. 'Mandatory Screening for Hepatitis B', 25 October.

Henry, M., 1996. 'Reducing Abortion Rates', *The Irish Times*, 27 April.

Holmquist, K., 1994. 'Tinkering with the gene genie?', *The Irish Times*, 8 January.

The Medical Council, 1994. *A Guide to Ethical Conduct and Behaviour and to Fitness to Practise*, (4th edn), Dublin

Chapter 18

Genetic screening and testing
A moral map

ROGEER HOEDEMAEKERS
Department of Ethics, Philosophy and History of Medicine
Catholic University of Nijmegen
The Netherlands

1.   INTRODUCTION

At a time when mapping of the human genome is considered an important objective in scientific research, mapping of the moral debates in the domain of genetic screening and testing in a health care setting may also be a worthwhile pursuit. A descriptive analysis of moral arguments as well as an inventory of the most important pragmatic considerations should give greater insight in the moral dimension of this medical practice. This kind of research seems worthwhile before a critical moral assessment of the practice of the genetic screening and development of a normative framework for this domain is undertaken. Such an inventory of moral and pragmatic arguments will also facilitate location and further exploration of possible underlying complicating factors, issues or disagreements which may prevent (further) growth towards consensus.

Various assumptions underlie this survey. The first is that moral dilemmas posed by genetic screening and testing cannot always be adequately solved by an appeal to the four important principles respect for autonomy, beneficence, non-maleficence and justice. This is due to other factors that may influence the debate. Detection of underlying beliefs, moral theories, assumptions and pragmatic considerations may help solve some of the dilemmas. A second assumption is that some disagreements and conflicts have not been resolved because they can be traced back to underlying controversies, for example

R. Chadwick et al. (eds.), The Ethics of Genetic Screening, 207–230.

about the goals of medicine, or confusion about the meaning and usage of important concepts. The third assumption is that difficulties are created because arguments can be used by different actors for different purposes. Arguments justifying or supporting a specific procedure or action at the macro-level of health or government policy, for example, cannot always be simply transposed to the micro-level where actions of individuals are assessed morally.

The aim of this paper, then, is to inventorise the various moral and pragmatic arguments, to detect the underlying values, assumptions and beliefs and to uncover possible underlying disagreements or conflicts. These underlying possible conflicts and confusions should be addressed before proposals are made to solve certain issues.

1.1   Method

The typology developed below is based on 20 publications on genetic screening and testing in the United states, United Kingdom and The Netherlands. These publications, listed in the bibliography, cover a period of more than 25 years (1968 to 1994), and have been selected because they were thought to summarise or discuss all the major controversial issues with regard to genetic screening and testing. The result is probably not a full account of all the arguments found in the extensive literature in this field, but it will give a good idea of how the important issues are approached and solved in these countries.

The various issues have been grouped under five different headings corresponding to the various consecutive stages of a genetic screening and/or testing programme. A first group of issues is connected with the selection of the condition and the best method of prevention. It includes consideration of aims and benefits as well as consideration of potential harm and risks of screening for a particular condition, a restrictive list, severity of the condition and incidence and/or prevalence of the condition. A second group is linked with issues concerning access: the choice of the target group, the best time (and setting) for genetic screening or testing and consideration of preconditions. A third group, procedural issues, includes education, voluntary or mandatory participation, strategies to enhance compliance and adequate screening procedures, including quality assessment, costs, and funding. A fourth group is related to counselling procedures and comprises discussions about non-directive or directive counselling, withholding information, the right not to know and disclosure to relatives. A fifth group of issues is connected with storage, control and use of genetic data. Use of genetic information by insurers and employers as well as use of the information for preventive measures or intervention are discussed here. Therefore issues concerning selective abortion are included here.

Many arguments used to solve the various issues in genetic screening and testing can be traced back to important values. Generally speaking, values can be described as ideals, experiences or states of affairs pursued by individuals, organisations and communities. For individuals, for example, health and well-being are important values, for organisations efficiency and/or profit-maximisation, for health or government authorities justice (adequate health care for all) may be an important value.

Some values, such as happiness or health do not seem to serve a further purpose because they are valuable in their own right. Other values can be seen as instrumental in achieving these 'basic' values. Efficiency, feasibility and profit-maximisation seem to be such values. It may be reasoned, for example, that efficiency may contribute to profit-maximisation, which in its turn may contribute to greater happiness. Inclusion in a more overarching category, and tracing back argumentations to these 'basic' values like well-being and happiness would require an extensive theoretical framework, but would not be very clarifying, however. For the purpose of this paper it seems sufficient to link arguments to particular values (moral and non-moral) and underlying beliefs or assumptions.

Values and ideals can be pursued by various courses of action and decisions. Actions and decisions always take place in a specific economic, political, legal, ideological and technological context, and they can be judged from various viewpoints. When considered from an economical perspective, a course of action is judged by important economic values and principles. When judged from a religious viewpoint, actions are judged by important religious values, ideals or rules. They may also be judged from a pragmatic viewpoint. This means that actions are judged by their implications, their effectiveness and political or legal feasibility, for example. Actions and decisions can also be judged from a moral point of view.

Moral judgements involve the application of moral norms and rules, which can often be traced back to important principles. These principles are closely connected with important values, such as autonomy, well-being and justice. Various ethical theories try to explain why it is good or right to pursue particular values, and values that according to these theories ought to be realised may be called moral values. Arguments that can be traced back to these moral values will be called moral arguments.

Usually two types of ethical theory are distinguished: deontological and consequentialist (teleological). Consequentialist theories judge actions as right or wrong by their consequences, by the benefit or the good they may bring to individual and community. Deontological theories do not judge actions by some good they produce, but they take the nature of the action and motives from which they are performed into account. Deontological and teleological reasoning may lead to conflicting conclusions. However, as noted earlier, actions and decisions are seldom judged from a moral point of view only. They can be judged from an economic, religious and pragmatic perspective. This means that these different viewpoints may clash and create dilemmas. It seems necessary to locate these conflicts. Sometimes pragmatic considerations may help solve these dilemmas, but as solutions will seldom be acceptable for all parties involved, they may deserve further scrutiny.

These considerations have led to the development of a kind of (moral) linkage map. For each issue of each consecutive stage of the genetic screening and testing procedure the major arguments have been listed and linked to underlying (moral and non-moral) values, principles, assumptions and beliefs.[1] Wherever necessary, a distinction has been made between deontological, teleological and more pragmatic arguments. In the concluding part of this paper some unsolved conflicts and difficulties will be briefly discussed.

2.   PLANNING AND DESIGNING OF SCREENING PROGRAMME

Two major types of genetic screening and testing can be distinguished.[2] Presymptomatic genetic screening may give information about conditions or susceptibilities threatening the future health of the screenee him- or herself. Genetic screening and testing can also yield information about risks or conditions that may seriously affect the health of future children and is usually performed for reproductive reasons. In addition, genetic screening and testing can also be used by insurers and employers to detect health risks. Issues related with these last two types of screening and testing will be included in the following.

2.1   Aims and benefits

It is not always possible to distinguish between aims and benefits, as aims are often described in terms of expected benefits. In the formulation of aims and benefits health authorities usually describe the expected effects of the screening or testing programme. Genetic screening or testing may bring benefits at macro-, meso- and micro-level. Community benefits are: reduction of the incidence of disease, improvement of public health and improvement of health of future generations. A further perceived benefit is reduction of the burden of disease for society: genetic screening can lead to a reduction of health care costs.

Genetic screening may also benefit health care organisations and private corporations. Insurance companies and employers foresee benefits. Employers may wish to know if their employees are susceptible to a work-related illness or an illness which may impair work performance in order to enhance efficiency and, consequently, their profit. Insurers may wish to undertake genetic screening to better calculate the risk an individual has to develop a disease or they may wish to avoid adverse selection, which may reduce their profit. For an increasing number of private corporations providing diagnostic genetic tests, profit-maximisation is an important benefit. Genetic screening or testing also brings benefit to health care providers, especially counsellors and specialists who may provide more accurate counselling, thus enhancing efficiency and reducing costs. Also timely intervention and avoidance of unnecessary diagnostic tests and expenses are perceived as benefits of genetic tests. It may help reduce health care costs at an institutional and community level.

For individuals major benefits envisaged are enhancement of health, (psycho-social) well-being, and autonomy. It may help reduce or alleviate (future) physical suffering of the individual screened, his or her family and it may help avoid future suffering of children. Genetic information can also bring greater psychological well-being, for example by offering reassurance in case of a negative test result, by enabling procreation with less anxiety, or by helping people to anticipate future problems and complications. Genetic information is also said to offer individuals a certain measure of control and to help an individual make important life choices. This objective, enhancing autonomy, is considered especially important in connection with carrier screening and pre-natal screening programmes.

Besides the benefits already mentioned individuals may wish to undergo a genetic test for health insurance purposes. This may bring economic advantage in case the health insurer is not informed.

2.2   Potential harms and risks

Health care authorities are usually aware that genetic screening does not only offer benefits to individual, corporation and community, but also generates dangers and risks. Recommendations to avoid or reduce these potential harms and risks usually aim at prevention of potential harm for individuals, but potential harm is also perceived at the meso and macro-level.

Major harmful effects perceived at the community level are: discrimination of populations where a certain hereditary disease is found frequently, generation of social pressure to participate in genetic screening programmes, and creation of new societal norms, for example a greater tendency to avoid the birth of handicapped children, or a greater tendency to hold parents responsible for the suffering of their children.

At the meso-level negative economic effects are perceived: genetic screening and testing may lead to early medicalisation and, hence, to an increase in health care costs. Widespread commercial genetic screening and testing may contribute to undesirable forms of social pressure to undergo genetic testing.

Perceived harms for individuals are mainly connected with the values of psychological well-being and autonomy: a screening or testing offer may create anxiety or confusion. Non-participation may lead to feelings of guilt and regret. In case of pre-natal screening there is awareness that genetic testing may cause physical damage to mother and future child. It may further lead to difficult and agonising choices and stress. It may also give the person screened a false sense of security because other potentially harmful genetic mutations may go undetected or because of laboratory errors. A side-effect of pre-natal genetic screening, information about non-paternity, may have disruptive effects on the family. Failure to keep up a change of life style after learning about a genetic susceptibility to disease may cause stress and lower self-esteem. Genetic screening of children may have a harmful effect on the children screened: it may lead to greater parental concern, which may have a negative effect on the psycho-social functioning of the child. Knowledge of a late-onset disease may also create great stress in the child. Psychological damage may also result from self-stigmatisation and stigmatisation by family members.

Threats to autonomy are also perceived. Family members of persons screened may not wish to know the results, but may learn inadvertently about the results. This may compromise free choice. Also the (future) autonomy of the child may be threatened when his or her right not to know is violated.

2.3   Severity

There is general, but no complete agreement on the requirement that the condition to be screened for should be serious. The approach is essentially teleological: the expected burden of the condition is assessed, e.g., life expectancy, degree of incapacitation, age of onset are considered. Views of health professionals and professional organisations play an important role. More pragmatic arguments used to leave the selection of the condition to individuals themselves are: the burden of a condition is difficult to predict, there is often little certainty about prognosis, the psychological burden is difficult to assess and it is difficult to draw a line between the burden of a condition and mere inconvenience.

2.4   Restrictive list

Especially in connection with pre-natal diagnosis there has been discussion about a restrictive list. The general policy is that there should be no pre-natal screening for conditions not associated with a great physical, psychological and/or economical burden for the prospective parents and that there should be no screening for non-medical reasons. (e.g. sex-selection). Arguments supporting the development of such a list are usually based on the respect for unborn human life and fear that parents decide on termination of pregnancy for trivial reasons. Respect for unborn life demands that the decisions are not wholly left to the parents. A pragmatic argument for such a list is that it would make genetic counselling easier. Also the slippery slope argument is used: if no lines are drawn, the reasons for termination of pregnancy may easily become more trivial.

Most counter-arguments at policy level are based on pragmatic considerations as well, and reflect a wish to leave decision-making to individuals. A restrictive list would be quickly outdated, it would diminish flexibility needed in individual cases, many would not accept such a list, some diseases have a wide range of severity and an emergency situation would be difficult to assess for outsiders. Besides, if abortion on social indication is legally possible, it is difficult to prevent abortion for medical reasons.

2.5   Prevalence and incidence

There is a tendency to offer genetic screening and testing for conditions that are (more) common. Priorities may also be based on the size of the (sub) population affected. It has been observed, however, that this would lead to unfair treatment because rare hereditary conditions would be neglected.

3.   ACCESS

Four issues can be discussed here: selection of the target group, selection of the best time (and setting) of screening, preconditions limiting access and enhancement of compliance. Most of the arguments and recommendations are presented at health policy level.

3.1   Selection of target group

There are several possibilities: genetic screening can be offered to all, to a high risk group only, as part of regular medical practice, at individual request or by referral. Recently it has become possible to have a genetic test provided by commercial firms directly to persons who are interested.

For genetic screening of the total population and groups with an increased risk prevention or reduction of the total burden of a disease is an important consideration and the best choice usually depends on the best balance of (expected) benefits over harms. However, benefits and/or potential harms are difficult to determine for each individual. Also, harms and risks may be unequally spread in a population. Fairness also underlies the argument that genetic screening should be offered to all, not only to a limited group.

Many pragmatic considerations of efficiency, cost-effectiveness and available resources determine screening policy. Cost-effectiveness is influenced by the degree of acceptability of the programme for the target group, and compliance numbers. On the other hand, a large demand may overwhelm existing screening, counselling or health care facilities. Efficiency may be enhanced by screening for diseases with high prevalence and/or incidence, by organisational and administrative efficiency and reimbursement of screening and testing.

3.2   Time of screening

Here, too, balancing of benefits and potential harms plays an important role in order to decide whether pregnant mothers, newborns, young children, adolescents, or adults constitute the best target group. Pilot studies are often recommended here. For newborn screening for treatable conditions, for example, the benefit of early treatment or care is an important consideration. For untreatable diseases potential harms like the burden of knowledge, risk of misinformation and misinterpretation, anxiety and self-stigmatisation must be weighed. For (carrier) screening for recessive hereditary conditions an important consideration is that before conception there are more options, which enhance autonomy. Pragmatic considerations also play a role. For example, the efficiency of a genetic screening programme might be decreased if genetic information is forgotten. On the other hand, greater efficiency is expected in case of pre-natal screening, because the information is immediately relevant and prospective parents are more motivated. These advantages are weighed against the disadvantage of more limited options, or the fact that difficult decisions have to be taken in a short time. A pragmatic argument against pre-natal screening is that women who are pregnant may come when pregnancy is already advanced, so that selective abortion may not be possible within the time-limits of the law.

3.3   Preconditions

One precondition is found in connection with pre-natal screening: parents who wish to undergo pre-natal testing should be willing to have an abortion in case of an affected fetus. An argument in support of this precondition is that genetic testing is costly, and a wish to avoid possible harm underlies the argument that the test-procedure poses a risk for the fetus. Arguments against this precondition are based on respect for autonomy: preconditions do not fit in with voluntary participation, genetic information may enhance options and it creates unacceptable pressure on the woman screened. The future autonomy of the child is threatened if parents decide not to terminate pregnancy in case of an affected fetus. Preconditions are also felt to be unjust, because they are discriminating. Fairness also underlies the observation that it is unreasonable to expect prospective parents to have made up their mind before testing. A more pragmatic consideration is that nothing can be done if parents change their minds about termination of pregnancy.

Another precondition is that prospective screenees should give prior consent that genetic information of interest to other family members will be passed on. One argument in defence of this position is based on the wish to prevent unnecessary harm. Counter-arguments are that this attitude is paternalistic, limiting free choice and a more pragmatic attitude is revealed in the argument that this may deter individuals from taking the test, thus affecting efficiency.

4.   PROCEDURAL ISSUES

Several issues fall into this category: voluntary or mandatory participation, education and compliance, quality control, cost-containment and funding. Most arguments have been presented at health policy level.

4.1   Participation

In the 1970s mandatory screening was defended on the ground that it was the task of public authorities to protect those who could not protect themselves. Another important argument was effectiveness: 100% participation was not possible in voluntary genetic screening programmes. Mandatory genetic screening was also said to be easier to organise and implement. Future health underlies the argument that mandatory screening leads to a healthy gene pool, and distributive justice forms the basis of another argument: it is unfair to put the burden of avoidable disease on others.

However, mandatory screening programmes were increasingly criticised, the principal objection being that it compromised free choice, invaded privacy and bodily integrity. More emphasis was laid on education and voluntary decision-making, involvement and support. Especially mandatory pre-natal screening was felt to violate parental decision-making in reproductive matters. Pragmatic arguments are also found here: an argument against the unfair burden-to-society argument is that no extra burden is created if the individual takes the (financial) consequences, in case (s)he refuses, but an objection raised against this argument is, again, based on justice: it would be unfair to an affected child to wholly put the burden of care on the parents by refusing any help.

4.2   Education

As consensus grew about the voluntariness of genetic screening and testing, more and more emphasis was put on pre-test information and counselling. The most important requirement for education is that it is objective, non-directive and includes a discussion of benefits as well as potential harm and risks. Complete information is considered a basic requirement for autonomous decision-making and free choice. Threats to free choice have been noted, notably various forms of directiveness and social pressure. Education is also thought to diminish societal dangers, such as stigmatisation and discrimination. It may promote well-being by removing unfounded anxieties. It also enhances efficiency: adequate pre-test information will reduce the need for extensive genetic counselling. In addition, fairness requires that information about screening and testing possibilities should be offered to all.

4.3   Compliance

Strategies aiming at (greater) compliance can be based on considerations of efficiency, cost-effectiveness, enhancement of well-being and reduction of suffering. Five different strategies have been proposed. Education, emphasising the benefits, is an important strategy, the assumption being that the more individuals know, the more likely they will comply. Involvement of community leaders is a second approach, which may be based on the value of autonomy (the target group should also be given a say) or on more pragmatic considerations (community leaders are thought to play an important role in positively influencing attitudes of the community members). Sometimes forms of persuasion are considered. Persuasion is objected to because of the pragmatic consideration that it may not be effective and because persuasion is not consistent with the principle of autonomous decision-making. A pragmatic consideration is that even in case of persuasion, not all individuals will conform.

A fourth strategy is to select the best setting for screening. Efficiency and compliance is supposed to increase if genetic screening forms part of ordinary medical practice and if the approach is more personal.

A fifth strategy, somewhat suggestive of a market-oriented approach, is to use the results of pilot studies to increase compliance by presenting the attitudes of the majority.

4.4   Quality control

Great care is usually taken to minimise harms and risks of the test itself such as laboratory errors and misinterpretation of test results. Besides prevention of harm distributive justice underlies decision-making regarding the sensitivity and specificity of a genetic test. Especially the benefits and potential harm for false positives and false negatives are weighed as the benefits and harms are to be distributed equitably. High sensitivity (and consequently few false negatives) is usually recommended, the principal reason being that the consequences of false negatives are serious, especially in case of pre-natal genetic testing.

4.5   Funding

Participation will be minimal if the costs of screening and counselling are not reimbursed. In case the test (and/or counselling) is not reimbursed, questions of justice have been put forward, because genetic screening and testing would then be possible only for the well-to-do. Sometimes the funding problem is put in the wider perspective of fair allocation of health care resources: financing of genetic screening should be considered with a view to other health care needs.

5.   COUNSELLING ISSUES

Four important questions are often discussed: is the principle of non-directiveness always justified? Is the counsellor allowed to withhold genetic information? Do individuals have a right not to know and is a counsellor allowed to pass on important genetic information to relatives, in case the person screened refuses to give consent? Most counselling issues are found at the meso- and micro-level.

From the very beginning of genetic screening and testing the requirement to provide adequate pre- and post-test counselling and support has been put forward, especially for screening in a reproductive context. Because of the potential individual harms informed consent was thought to be very important. Also respect for autonomy was an important argument: prospective screenees should not be exposed to dangers they were not aware of. Later a more pragmatic argument appeared, especially in the U.S.: negligent counselling could lead to malpractice suits or other sanctions.

5.1   Non-directiveness

Awareness of past abuses of hereditary information led to a strong emphasis on non-directive counselling. A further argument in support of this approach is based on respect for autonomy (reproductive decisions involve highly personal value preferences). From the beginning, however, non-directiveness has also been presented as an ideal that can hardly be achieved. Underlying this ambiguity may be the fact that non-directiveness is not consistent with the usual (more directive) approach of health professionals reflecting the more paternalistic attitude of wishing to protect patients from harm. A non-directive approach may also be unhelpful at a time when advice is urgently needed.

5.2 Withholding information

This has been defended on the grounds of efficiency: too much information might frighten potential screenees out of being tested. Genetic information was also withheld because it was thought to lead to considerable harm, possibly to suicide in case of serious disease, because the screenee might not be able to cope with the information. Objections to this paternalistic attitude were that medical professionals overstep their roles if they try to judge psychological damage and that doctors do not know their patients well enough to make decisions for them.

At present much importance is attached to autonomous decision-making and informed consent. Withholding information is only acceptable in case the information does great harm, especially to third parties. This also applies to unexpected findings, such as non-paternity. Respect for autonomy and the wish to avoid great harm are here in conflict. Pragmatic solutions for this dilemma are to obtain advance consent about what information should be provided. A counter-argument is that this might lead people to the decision not to participate, which might be to their disadvantage. Another pragmatic argument not to withhold genetic information is that it may diminish trust in health professionals.

5.3 The right not to know

Growing awareness that possessing genetic information may be harmful has led to increasing attention for the right not to know, which can be seen as a new aspect of the right to protection of the personal sphere. Violation of the right not to know would also not be consistent with the principle of respect for autonomy. The right not to know has also been based on the principle of avoiding harm: a person should not be forced to live with the burden of knowledge against his/her wishes. It has been disputed on the basis of distributive justice: is not-knowing acceptable, if ignorance will produce a child that will be a burden on health resources? The right not to know may also harm others who might profit from the information. An appeal to solidarity underlies the argument that the right not to know might be given up out of a sense of responsibility towards others.

At present the right not to know is given much weight but is seldom seen as absolute as the genetic information can also be of great benefit to others. Balancing of benefits and burdens is suggested in such cases, especially with regard to children and incompetent persons.

5.4   Disclosure to relatives

Genetic information may be of great value to relatives and passing this information on to relatives may raise difficulties in case the screenee refuses consent. This dilemma is created because two important values are in conflict: protection of the personal sphere and avoidance or reduction of suffering. There is great reluctance to disregard the right to personal privacy. Efficiency considerations play an important role as well: participation would be less if confidentiality of data could not be guaranteed, and it would diminish trust and confidence in the medical profession, leading to a less effective medical service. A third argument is based on fear that the exception will become the rule (slippery slope argument) and a fourth argument is that a legal duty to inform is difficult to enforce.

A widely accepted conviction is that passing on genetic information to others is an obligation of the person screened. Sometimes a mild form of persuasion is allowed in order to prevent serious harm to relatives. On the whole, however, priority is given to protection of personal sphere.

6.   USE OF GENETIC INFORMATION

Genetic information is thought to be of great value to individuals, who can make important life and reproductive choices on the basis of this information. Therefore issues linked with selective abortion are discussed here. Other issues are storage and control of genetic data and use of genetic information and biomaterial by third parties.

6.1 Storage and control of data

Benefits of storing genetic data and biomaterial are that this may prevent further extensive testing, facilitate future diagnosis and counselling of relatives, and contribute to greater accuracy in counselling. A more pragmatic argument is that it may help health professionals to fight liability charges.

Potential harms of storage are pointed out frequently. Invasion of privacy, misuse of information, a fear of control of reproduction by society, inaccuracy and possible mistakes when handling the genetic data.

Safeguards have been formulated for proper preservation of privacy. Some safeguards have been objected to. For example, the possibility to inspect the information may harm the screenee. Access to the information should also be limited because the privacy of other persons may be harmed. A pragmatic objection is that safeguards are difficult to enforce and that anonymisation is difficult to achieve. For storage of biomaterial safeguards have also been developed in order to protect the right to privacy, bodily integrity and to maximise individual autonomy (e.g., agreements on disclosure of further information, or anonymisation of biomaterial for further research).

6.2 Use of genetic information

Two important groups of potential users are insurers and employers. Insurers have an economical interest, because they believe genetic information leads to better assessment of risks. Individuals may be harmed because premiums are raised or because health or life insurance is refused. This is felt to be unfair. On the other hand, individuals who possess genetic information which the insurance company does not have may have an unfair disadvantage as they may profit from this by getting very high coverage. This may be to the disadvantage of other policy-holders who have to pay higher premiums. Benefits and burdens are not distributed fairly in this case. Furthermore, harm may be caused by misinterpretation of genetic data, and health professionals fear that use of genetic information by insurers may lead to decreased participation, thus reducing efficiency.

Individual autonomy may be violated because the individual is usually in a dependent position when genetic information is asked, for example because (s)he needs the insurance for a mortgage or health insurance. Privacy is invaded because the information may reveal information about relatives.

Employers may want to test prospective employees for genetic information which may reveal that working conditions damage their health. Prevention of harm seems to be an argument here, but there may also be financial considerations: genetic screening may lead to a reduction of costs and an increase of efficiency by exclusion of workers at risk. Screening is also said to be to the advantage of the community as it leads to a reduction of costs for occupational disease and, consequently, of health care.

Harm may be caused to the individual because workplace screening may confront workers with information they do not wish to know, and it may lead to exclusion of workers and not to improvement of workplace conditions. Loss of work because of genetic information is also discriminating and felt to be unfair.

6.3   Selective abortion

In the 1970s and early 1980s abortion legislation was introduced and selective abortion became less and less a public issue. It is realised that it is still controversial and raises a number of moral and social questions, but with increasing importance attached to autonomy and self-determination it has become more and more an issue to be resolved by individuals. Pluralism of values has been recognised and abortion has become a matter for the prospective parents, who should take a decision fitting in with their personal beliefs, values, convictions and life situation. Thus it could become a new form of prevention of disease, although it was sometimes considered a kind of interim solution for untreatable diseases, useful until therapy was developed. A tendency has grown to consider (pre-natal) genetic screening and selective abortion as separate issues, the latter to be solved by the parents.

Thus at policy level a rather pragmatic stance has been taken on the issue of selective abortion. Teleological arguments prevail. Selective abortion is thought to contribute to parental well-being because it prevents (serious) suffering for both parents and child. It may also lead to saving of health care resources. Benefit for future generations has been perceived, as selective abortion may contribute to better genetic quality of the human race.

Counter-arguments are usually based on respect for unborn human life. For some world religions this respect for unborn human life is absolute. A less absolute stance is usually justified by pointing out that unborn human life deserves respect but not necessarily the same as humans born. Respect for unborn human life underlies the recommendation that abortion for trivial reasons is not morally acceptable. The unprotected fetus should not be exposed to the wish fulfilments of the prospective parents. Therefore authorities must define limits (see above, under prescriptive list). Health and government authorities frequently appeal to public opinion and point to statistical evidence that selective abortion for medical reasons is widely accepted.

At the meso-level many health care institutions also tend to follow the pragmatic approach. Many physicians, specialists and counsellors feel that they should not interfere with individual decision-making. Non-directive counselling became an important principle from the very beginning. Respect for parental autonomy was one important principle, fear of being accused of eugenics another. For many health care professionals selective abortion has become a rational solution to prevent suffering and a positive alternative for parents with an increased risk of having an affected fetus. The idea has spread among providers of pre-natal screening and counsellors that they are increasing parental well-being: raising a seriously handicapped child threatens future well-being of the prospective parents and selective abortion is a method of avoiding this.

At the individual level three sorts of arguments are often used: teleological arguments focusing on the welfare of the fetus, parents and family, deontological arguments, and some more pragmatic considerations. Three important arguments may be classified as teleological: the burden of the fetus argument, the burden of the family argument and the total burden argument.

As for the burden to the fetus: fundamental underlying assumptions are that termination of pregnancy is the lesser of two evils and that non-existence is preferable to a life full of suffering. An argument countering this assumption is that non-existence and a life full of suffering cannot be compared, the more so because there is usually some uncertainty about the extent of future suffering.

Well-being, often formulated in terms of quality of life is an important consideration, the basic assumption being that there should be a minimal quality of life for the fetus; if there are serious doubts about this, selective abortion is believed to be justified. The welfare and quality of life of the parents is also an important consideration as it is assumed that a seriously handicapped child usually causes much suffering for the parents. Frequently these two arguments are combined: the burden for the parent or family is added up to the burden of the future child. Termination of pregnancy is said to diminish the total burden of suffering. Sometimes this is formulated positively: selective abortion enables parents to have healthy children.

It may be pointed out here that the principle of respect for life can be used in two opposing ways. On the one hand it is used to completely reject selective abortion, because the life of fetus deserves full respect, whether it will lead a miserable life or not. On the other hand this argument is used to justify selective abortion: a child should not be condemned to a miserable death and respect for human life demands that it should not be subjected to (degrading) and serious suffering.

As pointed out above, the principle of respect for life (or sanctity of life) is closely tied up with views of the fetus as a person. Those supporting the sanctity of life argument state that a fetus should be treated as a person at all stages of development, others take a less absolute stance, and defend their position by a gradualistic approach, or consider a fetus as a potential person. It may be noted that the potential person argument also gives rise to two opposing views: A fetus is *only* a potential person that has not yet developed its full potential (the gradualistic approach), the other position is that a fetus *already* has the full potential of a person and therefore deserves full respect.

7.   DISCUSSION

Some important moral values can be linked with many different disputes. Especially enhancement of autonomy, health, well-being, prevention and reduction of suffering and privacy turn out to be important normative determinants influencing the debate. Other major but essentially non-moral determinants, are financial considerations, profit, efficiency, and feasibility, especially at the meso- and macro-level.

The prevailing approach is consequentialistic: results and effects of genetic screening at macro-, meso- and micro-level are considered and assessed. Expected benefits, dangers and risks of courses of action are compared and balanced. If there is a net benefit, genetic screening or testing programmes are thought to be morally acceptable. A major question, well-known in connection with a consequentialistic approach, is what criteria can be applied to weigh the various benefits and harms for individuals and society. This question has not been properly solved yet. A tendency noticeable is that health and government authorities leave it to the individuals concerned to weigh their benefits and harms, not only because they are in the best position to judge them, but often also because of pragmatic reasons. A consequence of this is that potentially negative societal consequences, for which scientific evidence is more difficult to produce, tend to receive relatively little attention.

In all the discussions linked with genetic screening there are various sources of conflict leading to moral dilemmas. There are conflicts between the two major theoretical approaches, teleological and deontological, between conflicting (moral and non-moral) values, between moral and more pragmatic considerations and there are conflicting underlying assumptions or beliefs.

An example of a conflict between a teleological and a deontological approach is apparent in the moral debate about selective abortion. Advocates of selective abortion point at the alleviation of suffering for future child, parents, family and society. Opponents of selective abortion adhere to a more deontological approach and point at the nature of the act and the obligation to respect unborn human life. At the macro-level an appeal to parental autonomy as well as various pragmatic arguments have been put forward to solve this dilemma.

Conflicting moral values can be found at various levels, for example in counselling, where respect for the private sphere and autonomy may clash with the prevention of suffering of third parties. Moral and non-moral values may also conflict: commercial providers of genetic tests face the dilemma of proper education about all the uncertainties and potential harms and the prospect of heavy losses because many individuals may then decide not to use the test. Underlying assumptions or beliefs may also come into conflict, the best illustration probably being the debate about the moral status of the fetus, between the belief that unborn human life deserves absolute respect and the belief that the moral status of the fetus is dependent on the stage of development.

Most of the moral dilemmas and conflicts have already been under discussion for a long time, though not properly solved. Other problems have so far received inadequate attention. One of these is that individuals, who are to take the ultimate decisions, often rely on the medical profession for information to make the right choice. This puts health professionals in a powerful position, the danger being that subtle forms of pressure are used, knowingly or not knowingly, to guide individuals towards the 'proper' decision. Involvement of community leaders and other strategies to enhance compliance may serve as illustrations here. We may probably assume that health professionals are be mainly motivated by paternalism and an urgent wish to do good. It is to be doubted, however, that this is uppermost in the minds of commercial providers of genetic test and services, whose main drive is probably to make profit. In both cases great care should be taken that pre-test information is objective and non-manipulative. Analysis of literature regarding existing large-scale genetic screening programmes and analysis of educational or promotional material about genetic tests should be undertaken to detect if subtle forms of coercion are used.

Because the medical profession plays a major role in education and counselling it is important to further analyse what sort of concepts of health and disease they rely on. Different health models may lead to different approaches regarding the goals of medicine. For example, if a holistic health concept is favoured, which includes physical, psychological and social well-being, it is understandable that enhancement of autonomy has gained such importance in the aims of screening. If, on the other hand, a more reductive health concept is favoured, which emphasises physical health, enhancement of autonomy seems less obvious as a goal of a medical practice.

An important recommendation for screening programmes is that the condition to be screened for must be serious. But what is a serious disease? Nowadays consensus is growing that value judgements play an important role in concepts of health and disease. Analysis of value judgements at different levels with regard to the concept of disease, genetic disease and genetic abnormality may be clarifying. In view of the fact that prevention and alleviation of suffering is a frequently stated goal of genetic screening and testing, judgements about normality and abnormality in the context of medical genetics may also be illuminating. Not every genetic or chromosomal abnormality results in abnormal degrees of suffering. And in how far do judgements about abnormal suffering between health professionals and prospective screenees diverge? Here may be another (and hidden) source of (subtle) coercion. Further analysis here may help answer the question who must decide on the introduction of a genetic screening programme.

Prevention and/or alleviation of suffering is an important goal of genetic screening and testing. Assessment of normal (acceptable) and abnormal (unacceptable) degrees of suffering are usually tied up with quality of life considerations. The advantages and disadvantages of the concept of quality of life in the moral debate about selective abortion, we suggest, also deserve further exploration. The more so because confusion, and a conflict of interests is possible between the parents and their quality of life and the unborn child and his or her future quality of life. Analysis is needed because professional, individual and societal judgements about the quality of life may be closely interwoven and may thus offer another possibility of manipulation.

Because of the many inherent uncertainties and dangers of genetic information further exploration of the value of this information for individuals also seems indicated. On the one hand the general impression is that genetic information is felt to be important. On the other hand it is frequently observed that this information is also difficult to understand and that misinterpretation is easy. We believe therefore, that insights gained from literature about risk, risk-perception and risk-interpretation might also contribute to further understanding of the concept of risk in medical genetics. The same applies to another key concept in medical genetics: increased risk. Introduction of a genetic screening programme should not only depend on the degree of suffering associated with a particular genetic abnormality, but also on the probability with which great suffering will occur. Here professional and individual perceptions of risk may diverge. What is the proper role of the medical profession here? Presenting a worst-case scenario? And what exactly is the meaning of the concept of increased risk in the total life-plan of a particular individual? The danger is that the increased risk presented by a genetic abnormality becomes isolated from an individual's life project and other risks threatening an individual's health or safety and is overemphasised.

ACKNOWLEDGEMENTS

I am grateful to the Commission of the European Communities for funding the Euroscreen project of which this paper forms a part. Part of this paper has been presented at the Euroscreen plenary in York, in November 1995. It has benefited from discussions with members of the Core-group of the Euroscreen projects, notably Prof. H. ten Have, Catholic University Nijmegen, The Netherlands and Prof. J. Husted, Aarhus University, Denmark, and members of the subgroup concerned with philosophical aspects of genetic screening: Prof. I. Pörn, University of Helsinki, Finland, Dr. K. Dierickx, Catholic University Leuven, Belgium and Dr. V. Launis, University of Turku, Finland.

NOTES

1 It should be noted that for the purpose of this paper it did not seem necessary to link each argument with a specific publication. Often the same arguments are used in various publications. Selecting one or two for reference purposes would probably not do justice to other publications and their contribution to the moral debate.
2 Although a distinction can be made between (large-scale) genetic screening and genetic testing, it does not seem really useful here to clearly distinguish between the two. Not only because often these two approaches are not clearly distinguished, but, more importantly, because many arguments can be used for genetic screening as well as for genetic testing.

REFERENCES

Andrews, L.B. et al (eds), 1994. *Assessing Genetic Risks, Implications for Health and Social Policy*, Institute of Medicine, National Academy Press, Washington D.C.
Bergsma, D, et al., 1974. *Ethical, Social and Legal Dimensions of Screening for Human Genetic Disease*, New York, London.

Council of Europe, 1994. Recommendations R(92)3 on Genetic Testing and Screening for Health Care Purposes, reprinted in, *Gezondheidsraad, Commissie Screening en Erfelijke en Aangeboren Aandoeningen, Genetische Screening*, pp. 123-128, Den Haag.

Powledge,T.M. and Fletcher, J., 1979. Guidelines for the Ethical, Social and Legal Issues in Pre-natal Diagnosis, (Genetics Research group of the Hastings Center, Institute of Society, Ethics and the Life Sciences), *The New England Journal of Medicine*, pp. 168-172.

Gezondheidsraad, 1977. *Advies inzake Genetic Counseling*, Rijswijk.

Gezondheidsraad, 1980. *Advies Inzake Ethiek van de Erfelijkheidsadvisering (Genetic Counselling)*, Den Haag.

Gezondheidsraad, 1989. *Erfelijkheid, Wetenschap en Maatschappij, Over de Mogelijkheden en Grenzen van Erfelijkheidsdiagnostiek en Gentherapie*, Den Haag.

Gezondheidsraad, 1994. *Commissie Screening Erfelijke en Aangeboren Aandoeningen, Genetische Screening*, Den Haag.

Harris, M., (ed), 1972. *Early Diagnosis of Human Genetic Defects*, Washington DC. in B. Hilton, et al. (eds), 1973. *Ethical Issues in Human Genetics, Genetic Counseling and the Use of Genetic Knowledge*, New York, London.

National Academy of Sciences, National Research Council, Committee for the Study of Inborn Errors of Metabolism, Genetic Screening, 1975. *Programs, Principles, and Research*, Washington.

Nuffield Council on Bioethics, 1993. *Genetic screening, Ethical Issues*, London.

President's Commission (for the study of Ethical Problems in Medicine and Biomedical and Behavioural Research), 1983. *Screening and Counselling for Genetic Conditions*, Washington DC.

Lappé, M. et al., 1972. 'Ethical and Social Issues in Screening for Genetic Disease, Research group on Ethical, Social and Legal Issues in Genetic Counseling and genetic Engineering, Institute of Life Sciences', *New England Journal of Medicine*, 286, pp. 1129-1132.

U.S. Congress, Office of Technology Assessment, Cystic Fibrosis and DNA Tests, 1992. *Implications of Carrier Screening*, U.S. Congress, Washington DC..

de Wert, G.M.W.R. and de Wachter, M.A.M., 1990. *Mag Ik Uw Genenpaspoort? Ethische Aspecten van Dragerschapsonderzoek bij de Voortplanting*, Baarn.

Wertz. D.C. and Fletcher, F., (eds), 1989. *Ethics and Human Genetics; a Cross Cultural Perspective*, Heidelberg, New York.

Wetenschappelijk Instituut voor het CDA, 1992. *Genen en Grenzen, een Christen-democratische Bijdrage aan de Discussie over de Gentechnologie*, Den Haag.

Wetenschappelijke Raad voor het Regeringsbeleid, 1988. *De Maatschappelijke Gevolgen van Erfelijkheidsonderzoek*, Den Haag.

Wilson, J.M.G.and Jungner,G., 1968. *Principles and Practice of Screening for Disease*, World Health Organization.

Table 18.1

	health	well-being	suffering/harm	autonomy	justice	bodily integrity	privacy	respect for life	efficiency	profit	feasibility	assessment problems	other
benefits community	*	*	*							*			
benefits institutions									*	*			
benefits individuals	*	*	*	*						*			
harm community			*	*	*								
harm institutions				*						*			
harm individual	*	*	*	*									
severity	*	*	*									*	
restrictive list	*	*	*	*				*	*		*	*	*
prevalence/incidence					*								
target group	*		*		*				*	*	*		*
time	*	*		*					*	*	*		
precondition	*		*	*	*				*	*	*		

Table 18.1 (cont.)

	health	well-being	suffering/ harm	autonomy	justice	bodily integrity	privacy	respect for life	efficiency	profit	feasibility	assessment problems	other
participation	*		*	*	*	*	*		*				*
education		*	*	*	*				*				
compliance	*	*	*	*					*		*		
quality control			*		*								
cost/funding					*				*	*			
non-directiveness		*	*	*									*
withholding information			*	*					*				*
right not to know			*	*	*		*						*
disclosure			*	*			*		*		*		*
storage/ control			*	*		*	*		*	*	*		*
use of information			*	*	*		*		*	*			
selective abortion	*	*	*	*				*		*	*	*	*

Chapter 19

The genetic testing of children

ANGUS CLARKE
Institute of Medical Genetics
University of Wales, Cardiff
Wales

1.   INTRODUCTION

The importance of genetic diseases in childhood has grown over the course of this century as that of infectious diseases and malnutrition has diminished, at least in developed countries.   Accordingly, many of the children now requiring medical attention are recognised as suffering from genetic conditions.   Making the diagnosis of a genetic disorder in an affected child is therefore a standard clinical activity. Establishing such a diagnosis has traditionally depended upon finding evidence of the disease process but is increasingly turning to genetic investigations that identify the underlying, genetic cause of the disorder. Such genetic tests can be carried out on any tissue and at any age, from conception onwards, because they do not depend upon the condition having manifested itself in any way. It is therefore possible to identify healthy children who will, in the future, develop a genetic disease; it is also possible to identify those who are healthy carriers of a disease which will never affect them but which may be of relevance to their own future children.

A sick child warrants full investigation to determine the cause of their illness, to ensure that the child is given the correct treatment and is managed appropriately; this may entail genetic tests. This does not mean, however, that it is always helpful to identify healthy children likely to develop or transmit genetic disease in the future. Even diagnosing an affected child as

R. Chadwick et al. (eds.), The Ethics of Genetic Screening, 231–247.

having a genetic disease, as opposed to any other type of disease, may have serious repercussions which require careful consideration as a family becomes aware of them.

The fact that the disease affecting a child is found to be genetic may have far-reaching consequences for the family. The nature and extent of these consequences will often depend upon the mode of inheritance of the condition, which will determine which other family members will be at risk of developing the condition themselves or of transmitting it to their children. For example, if it is transmitted as an autosomal recessive trait, then the child's present or future siblings may develop the same disease but it would be less likely for the child's own future children to do so. Information like this will often be unwelcome but it is generated simply by establishing the diagnosis in the affected individual and so it cannot be avoided. Furthermore, members of the family will sometimes find the information to be helpful if it allows them to avoid complications of the disease or to take the risk of it into account when planning future children.

When a child presents with symptoms or signs of an illness that may be genetic in origin, therefore, any test carried out to establish the diagnosis may have important consequences both for the child and for other members of the family. This will be true whether or not the techniques used for establishing the diagnosis are genetic. It will always be important for clinicians to recognise these possible consequences and to help the child and the family adjust to their new situation, once the diagnosis has been made. Indeed, this is one of the principal tasks of clinical geneticists and genetic counsellors.

A very different situation arises when a family or a health professional considers the question of testing a healthy child to find out about his or her genetic constitution. This could arise in at least three different contexts: i. predictive testing to see if the child will develop a disease recognised in another family member (parent, uncle, cousin, brother or sister ...), ii. carrier testing to see if the child may be at risk of having an affected child him or herself when older, or iii. a screening test unrelated to family history to identify children with increased susceptibility to some common disease such as ischaemic heart disease or diabetes. The scope of all these types of testing has increased dramatically over the past ten years, and will continue to increase as more genes are identified in the course of the Human Genome Project. These tests can be carried out on minute samples of any tissue from the body (even a mouthwash) and do not depend upon the child having developed the disease; they recognise the child's genotype and do not depend upon the phenotype (manifestations of the disease process). In this chapter, it is the issues raised by the genetic testing of healthy children that will be examined.

2. PREDICTIVE TESTING

When will it be helpful to carry out genetic testing to see if a healthy child will or will not develop a specific disease that runs in the family ?

This type of predictive genetic testing will clearly be appropriate if the child is at risk of complications of the disease that can be avoided or ameliorated if recognised early. For example, if a parent has familial adenomatous polyposis coli (FAP) - an autosomal dominant condition predisposing to cancer of the large bowel - their child will be at a 1 in 2 (50%) risk of developing the same condition. Surveillance for tumours in FAP is often started at 10-12 years of age, so predictive testing before then would be appropriate to distinguish those children actually at risk of tumours, and in whom surveillance would be useful, from those who could be spared the unpleasant annual colonoscopy.

It may also be helpful to the family as a whole to carry out predictive testing if a child is at risk of developing a serious illness in childhood. There may be an affected brother or sister, or perhaps an uncle or cousin, and the family may be looking for early signs of the same condition in another child; they may be in emotional agony, watching the child to see if he develops early signs of the same condition. For example, if a boy has a progressive, sex-linked disorder such as Duchenne or Becker muscular dystrophy, then his younger brother may be watched intently to see if he shows any of the early signs of muscle weakness. It can be very helpful to test the younger boy to find out definitely whether he does or does not have the condition. If he is affected then he is no worse off; if he is unaffected then the family will be spared their frequent misinterpretations of fatigue or cramps as early signs of the same condition.

In other contexts, however, where the disease does not usually present until well into adult life and where there is no useful medical intervention to be made, the predictive testing of a child could have damaging consequences. Before predictive testing became possible for Huntington's disease (HD) - a disease causing memory loss, personality change and a movement disorder - there was debate about the circumstances in which such testing should be offered. This was raised as a concern by Craufurd and Harris (1986), and a consensus decision not to test children was reached by the World Federation of Neurology (1989 and 1990) and the International Huntington's Association (a federation of lay support groups). This was reinforced by the considered opinion of others involved in predictive testing for HD (Bloch and Hayden, 1990).

The grounds for such caution in relation to predictive testing of healthy children for late-onset disorders are essentially fourfold:

1. If a child is tested then they lose the opportunity to make their own decision in the future as an autonomous adult.

2. When an adult is tested, they have control over the result and how it is used and disseminated; even the fact of their being tested is treated with strict confidentiality. A child who is tested forfeits confidentiality in respect of both the fact of being tested and the test result.

3. The family's knowledge of the test result could result in the child being treated differently, and this might lead to social or emotional problems in the child. A child destined to develop the disease may be brought up with lower expectations for their education and career or altered expectations for their future relationships. Such expectations could then become self-fulfilling prophecies, and could damage the self-esteem of the child at the deepest levels. Even a favourable result may have difficult emotional consequences; a child shown not to carry the family's disease gene may feel excluded from the inner circle of family concerns. Problems within the family may be particularly likely if the test result conflicts with the result anticipated by family consensus in the context of 'pre-selection' (Kessler, 1988).

Experience with predictive testing for HD in adults has shown that problems arise in a high proportion of cases. Most of these problems are not technical, laboratory problems but relate to counselling issues: the appropriateness of the test request, the attitudes of other family members or the disclosure of the test results (European Community Huntington's Disease Collaborative Study Group, 1993). Furthermore, problems are as likely after a favourable as an unfavourable result, although the adverse consequences may be delayed for some months especially when the result has been favourable (Codori and Brandt 1994; Huggins et al. 1992; Lawson et al., 1996; Tibben et al., 1992).

Particular problems could arise if more than one child in a sibship were tested, and they were given different results. This could easily lead to great difficulties in the relationships within the family, between the parents and children and among the children. The possibility of adverse consequences for insurance or employment in the future would also need to be considered, because an individual found to have the HD gene mutation might find him/herself to be excluded from a career or from health or life insurance as if already affected by the disease, despite being perfectly healthy at that stage. Given that the problems experienced in the testing of adults have arisen despite intensive prior counselling of the tested individual, and that an individual to be tested in early childhood would often not be able to participate in such counselling, then there are good grounds for caution in this area.

Genetic counselling has, for good reasons, developed an adherence to the view that genetic testing should be linked to counselling of the individual to be tested in advance of the test itself; the experience with testing for HD has reinforced this tenet of the profession. Testing children would run counter to this principle, and the potential additional problems anticipated if children were to be tested combine to make such predictive testing most unwise.

4. An additional reason for commending caution towards predictive testing in such circumstances is that a majority of adults at-risk of HD choose not to be tested. Whereas a clear majority of at-risk family members stated hypothetically that they would seek testing once it became available, in fact only 10-15% have done so; when faced with the option in reality, most have chosen to retain uncertainty (Ball et al., 1994; Craufurd et al., 1989; Bloch et al., 1992; Tyler et al., 1992). When a parent considers whether or not they wish their child to be tested for such a condition, they may respond as if they were confronted with the question hypothetically for themselves; the reality of the question from the perspective of the child may not have the same impact on the parent.

These considerations are not just abstract concerns, but the issues raised have had to be considered in many centres where requests for the predictive testing of children at risk of HD have been received from parents, adoption agencies and other social or legal bodies (European Community Huntington's Disease Collaborative Study Group, 1993; Morris et al., 1988; Morris et al., 1989). Furthermore, these issues do not arise just in relation to HD but also in relation to an increasing range of other disorders as well (Harper and Clarke, 1990).

Given the small number of adults who in practice choose to be tested for HD when the test is available and they are confronted by the issues, and given that the rights of the child as a future adult to both autonomy and confidentiality are abrogated by testing in childhood, it is clear why so many professionals and lay groups are opposed to the predictive genetic testing of healthy children for such disorders (Dalby, 1995). Of course the family (and perhaps the child if sufficiently mature) may be relieved if a favourable result is given *but* the result will often not be favourable, and even if it is favourable the child may still become a casualty of testing.

3.   PARENTAL RIGHTS

Surely, it may be said, it is a parent's right to insist upon testing his/her child, both because the parent may simply wish to know the child's genetic status and because the parent may be thought to have a duty to discover any fact relevant to the child's present or future welfare. A child's parents will usually have the best interests of their child at heart, and there is no reason to equate the seeking of genetic testing for a child with child abuse; so why not comply with the wishes of the parent? While the concept of a parental 'right to know' has been argued in North America (Pelias, 1991; Sharpe, 1993), although not by all parties (Clayton, 1995),this approach has been discarded in Britain. The Children Act of 1989 considers parents to have duties towards their children rather than rights over them (Montgomery, 1993; Montgomery, 1994). Decisions about the child's health care - including predictive genetic testing - would need to be made on the basis of the long-term best interests of the child. A health professional who considered that genetic testing would run counter to the child's interests, and should be deferred until the child is older, could therefore legitimately refuse to carry out such tests.

While the process of genetic counselling will usually resolve differences of opinion between professionals and parents, this will not always happen. If family and professionals disagree, whose opinion should prevail? In the short term it may be the professional who 'wins', but this is not likely to be the end of the matter. Except in extreme cases, it will usually be possible for the family to find another professional who is willing to comply with their request. If not, it may be possible for the family to obtain the testing, albeit under false pretences, through commercial channels - by sending a sample for analysis through the post, as if the sample came from an adult. This reinforces the importance of a sensitive and non-confrontational counselling process in advance of any decision about testing, which will usually lead to a consensus view on how to proceed.

4.   WHO IS A 'CHILD' ?

The discussion so far has begged the question of who is to count as a 'child'. It is clear that any young child - of pre-school age - will be unable to consider the full range of issues involved in genetic testing. It is equally clear that some 17 year-old individuals are as emotionally and intellectually mature as some 18 year-olds, or 15 year-olds as 16 year-olds. It would not be helpful to adopt a sharp age-related criterion to draw a line between childhood and adulthood. This would inevitably be arbitrary and indefensible.

It will be much more helpful to treat each child and each set of family circumstances as individual. It is recognised in British law that children under 16 years can be competent to make decisions about their health care and reproduction without parental involvement, depending upon the maturity of the 'child'. What will be important in the context of genetic testing is to examine three factors. First, it will be necessary to ensure that a request for testing a legal minor comes from the minor and not from other individuals or agencies. Second, it will be necessary to ensure that the minor is indeed sufficiently mature to appreciate the possible consequences of a favourable or an unfavourable test result, including the consequences for reproduction, career, insurance, self-esteem and relationships within the family. Thirdly, it will be important to find out why the request for testing has come now rather than in a few years, and why a test should be performed *now*, at this time rather than another. Given a satisfactory discussion and clarification on these points, genetic testing of minors may be appropriate.

An assessment approach has been developed for handling requests from adolescents for predictive testing for HD (Binedell et al., 1996). Work with children in other health care contexts - especially consent for surgery - has shown that the maturity of a child can be greatly influenced by relevant experience, and that chronological age is not an adequate guide for assessing competence to participate in decision-making (Alderson, 1993). In this chapter, therefore, when caution is expressed in relation to genetic testing in childhood, this does not imply the adherence to an arbitrary age-based definition of childhood; nor does it imply that competence is suddenly acquired or thrust upon a 'child' as s/he becomes a 'mature adult'. Children acquire the ability to contribute to the making of important decisions at different rates, and the extent to which a particular child can participate in a specific decision will depend upon many individual and context-specific factors that have to be considered in the counselling before any decision is made about testing in that case. The focus of this chapter, then, is on situations where all would agree that the child is too young to participate seriously in the making of an important decision. This defines an approach which can then be modified appropriately in the various grey areas that arise in practice.

5.   ADJUSTMENT TO GENETIC INFORMATION

It can be argued that a child will be able to adjust more readily to unwelcome genetic information than will an adolescent or young adult. Unfortunately, there is very little evidence on this point. We need more information about the different ways in which families have communicated information about genetic risk and genetic tests, and how the young people concerned have responded to the information. Indeed, it may be true instead that adolescents respond better to genetic information and test results if they are explicitly given control over the process of counselling and testing instead of simply being presented with the facts. This may serve to reinforce their self-esteem and to promote their coping abilities.

6.   CARRIER TESTING

The arguments deployed above could also be applied to the rather different context of the genetic testing of children to identify unaffected carriers of genetic disease, especially autosomal or sex-linked recessive diseases but also carriers of balanced, familial chromosomal rearrangements. In these circumstances, the test result would not be relevant to the health of the child at all, but could be relevant to the child's future reproductive decisions. A test result indicating that the child carried one of these conditions would mean that their future children would be at risk of being affected. In the case of an autosomal recessive disorder, the child's future children would be at risk if their partner was also a carrier of the same condition. If the child was a carrier for a sex-linked disease or for a balanced chromosomal translocation, then any of her future children would be at risk irrespective of the genetic constitution of their partner.

The arguments against carrier testing in childhood are similar to those given above in the context of predictive testing, but they are somewhat weaker. The abrogation of autonomy and confidentiality will occur in this context too, but might not be viewed as so serious as in predictive testing. Testing may result in alterations to the parents' expectations of the child, but their scope will be more restricted - being focused on issues of personal relationships and future reproduction. The other issues of genetic discrimination and the stigmatisation of carriers are real issues that must not be forgotten, because examples of such consequences have arisen in a number of different carrier screening programmes, but they are discussed in the chapter on screening for carriers of recessive disorders and they will not be rehearsed here.

In families where a condition is known to be transmitted through healthy female carriers, such as the sex-linked disease Duchenne muscular dystrophy, it is clear that girls are often brought up with powerful expectations about the pattern of their future relationships and reproduction being held by parents (especially by their mothers), and these expectations are frequently adopted by the daughters (Parsons and Atkinson, 1992; Parsons and Atkinson, 1993; Parsons and Clarke, 1993). Such patterns of family expectation arose before genetic testing was able to distinguish between those who did carry the disease in question and those who did not. With the development of accurate methods of carrier testing, families may consider clarification of a child's carrier status while she is still too young to be involved in the discussion. Is it reasonable to comply with such requests, or should the professionals act as gatekeepers to testing and be more cautious, seeking to protect the child's future rights? In autosomal recessive diseases such as cystic fibrosis, it is also known that the healthy sibs of affected children have many needs for information, counselling and support whether or not they turn out to be carriers (Fanos and Johnson 1995a; Fanos and Johnson, 1995b). In some families there are strong family expectations about genetic testing. Once again, should professionals comply with requests to test young children?

As in the context of predictive testing, an open discussion with the parent(s) or others requesting testing - reassuring them that testing will be available when the request for it comes from the child and at a time when it is relevant to the child - will often resolve the issue. The parents may have thought of testing the child as a way of discharging the duty of responsible parents to ensure that the child is tested. In fact, of course, testing a young child does not discharge the duties of either the parents or the professionals, it merely alters them. Whatever the test result - whether the child is a carrier or not - it will still be necessary for the parents to raise with the child the possibility of their being a carrier at an age when they can understand the significance of the question. Unless the child is forever kept unaware of the genetic condition that has occurred in the family and so is unaware of their potential carrier status, even a negative test result ('not a carrier') will have to be explained to the child when older. The professionals involved with testing a young child for carrier status will also not thereby have discharged their duties simply by giving the result to the family; there remains an obligation to ensure that the child is given the opportunity to discuss their result when older, thereby ensuring that they actually are given the result and that they understand its significance. It is all too easy for families to forget about such tests (especially if a crucial family member dies or becomes ill), or to misunderstand the significance of a test result.

Some families will express a very strong desire to test a child for carrier status despite a very full discussion. There may be a hidden agenda in some such cases, such as a desire to test paternity, but it may simply be a strong 'desire to know'. The family context will often be of parents who have lost a brother, sister, son or daughter with a serious and most distressing disease; unresolved grief in the parents may - very understandably - intrude into the decisions being made about the child. In my experience, it is not helpful for professionals to refuse to perform carrier tests under these circumstances. It is very reasonable to explore the motivation and reasoning underlying the parents' request, to indicate that testing will not resolve all the issues even if the child is shown not to be a carrier and to explain the possible difficulties that may arise from testing in childhood; but if the parents continue to insist on the child being tested then it may be appropriate to agree to this, because a persistent refusal generates antagonisms that may block the resolution of whatever factors are underlying the request. One approach to handling this would be to agree to test the child in principle but to insist upon a delay of at least a few weeks, after which the family has to re-initiate contact with the clinic; the decision is put entirely in their hands. This may give the family a chance to reassess their decision without the fear that the test will be denied them, so that for the first time they become free to examine all the factors for and against testing and then to make up their own minds.

7.   SUSCEPTIBILITY SCREENING FOR COMMON DISEASES

Experience with screening children for genetic susceptibility to common diseases is limited, but the experience that has been accumulated amounts to a cautionary tale. A report from Bergman and Stamm some 30 years ago introduced the term 'cardiac nondisease' to describe the emotional and physical effects of inappropriately labelling healthy children who had benign cardiac murmurs as being sick (Bergman and Stamm, 1967). In the newborn period, a Swedish programme of screening to identify infants affected by alpha1-antitrypsin deficiency was stopped because of damaging social consequences. Affected children are more susceptible to lung disease in adult life and should be protected from exposure to cigarette smoke, but the fathers of the identified infants actually smoked more heavily than the fathers of control infants (Thelin and McNeil, 1985). In USA, children found to have above average serum levels of cholesterol have sometimes been subjected to excessively restrictive diets, even leading to growth failure from malnutrition (Lifshitz and Moses, 1989; Newman et al., 1990), and sometimes their possible special health needs have been ignored (Bachman et al., 1993).

The lessons I draw from these reports are i)that programmes of screening for genetic disease in childhood should be piloted and considered very carefully before being adopted on a wide scale, and ii) that labelling healthy children as having a faulty gene could well lead to similar problems to those reported by Bergman and Stamm, and should not ever be undertaken lightly.

8. AN 'IDEAL' APPROACH TO GENETIC INFORMATION IN CHILDHOOD: THE RELEVANCE OF ADOPTION

While requests to carry out predictive genetic tests for late-onset diseases on healthy children rouse fears in many professionals that such testing will remove the child's future rights to autonomy and confidentiality and may be frankly damaging to the child's welfare, these fears are not so marked in relation to carrier testing of purely reproductive significance to the future adult. As discussed above, therefore, it may be reasonable to accede to a family's request for carrier testing after a full exploration of the reasons underlying the request and after counselling about the possible implications of the testing for the child; the consequences for the emotional welfare of the family of a persistent refusal to test may outweigh the reasonable concerns about testing. For predictive testing, in contrast, my judgement is that the protection of the child's future rights and interests will permit their testing only in the most exceptional circumstances. But the preferred course of action in both circumstances, relating to predictive and carrier testing, would be to offer full counselling but to defer the testing until the child could participate and make up his or her own mind when older. So when a family requests genetic testing of their child, what would be the optimal set of arrangements to make with them in relation to genetic testing for the child in the future ?

A guide to the best course of action in these circumstances may come from the general advice offered in relation to telling adopted children about their biological parents. It would be a mistake ever to lie to or to mislead children, and the best policy is to give information to children at the rate at which they seek it. In this sense, it has much in common with 'breaking bad news' in the context of cancer diagnoses (Buckman, 1992). Children can be introduced to the idea that they have been adopted - indeed, they have been 'specially chosen' - from a very early age. But they are not usually given much information about their biological parents until they are older. Adopted children are not permitted to seek out their biological parents until they are adult. A similar approach can be used with genetic information; children can be brought up in the knowledge that they may share certain genes with other family members, and that they will be able to find out more about this once they are older. The parallel is very close, and of course genes are involved in both situations. If both the parents and the professionals are to be reassured that they are discharging their responsibilities properly, then there is clearly a need for someone to ensure that the child is offered genetic counselling at an appropriate time in the future. The family may 'forget' about the genetic disease, especially if the affected individual in the family died before the 'child' was born. This is particularly likely to occur if a parent or another crucial family co-ordinator dies, or if the parents separate or divorce. But there is often no clear mechanism for the health professionals to 'remember' to contact the child when s/he is older, and in any case the family may well have moved house or country in the meantime.

This is not an argument for testing children in childhood, because the need for genetic counselling remains even once the test has been carried out - it could even be seen as more important because the family may have misunderstandings and may not appreciate that further discussion is required if they think that the testing process has been completed. Even if the family was perfectly informed 10 years previously, testing and the interpretation of test results may have developed in the meantime. Rather, this is an argument for the continued existence of regional genetic counselling services. These services should be adequately supported so that they can maintain infrequent but regular contact with families in which individuals may wish in the future to be offered genetic counselling and perhaps testing. This amounts to an argument for the funding of genetic services to enable them to maintain genetic registers - lists of patients who can be offered contact in the future for genetic counselling or, in a wider context, for surveillance for the possible complications of genetic disorders.

At the present, the transmission of genetic information - knowledge about a disease that has cropped up in the family as well as of test results and the appropriateness of genetic counselling for other family members in the future - relies upon three systems: family knowledge and willingness to transmit relevant information, the general medical practitioner or 'family doctor', and genetic counselling services. With the increasing complexity of genetic information, the increase in family instability and the more frequent changes in 'family doctor' that accompany social mobility, there is a good argument for strengthening the role of genetic services to compensate for changes in the other two systems. This is not a call for a new institutional framework but a plea for the more adequate support of genetics services within the health service to permit them to carry out more systematically a role that has often been thrust upon them but for which they have never been adequately resourced. This type of genetic register would be completely voluntary and would require the active co-operation of the new legal burden on genetic services never to lose contact with their clients (Hoffmann and Wulfsberg, 1995).

9.   EVIDENCE AND THE FUTURE

The Clinical Genetics Society in Britain established a working party to examine the range of attitudes and practices in relation to the genetic testing of children, and it examined in particular the problematic areas of predictive and carrier testing in childhood. The report of the Working Party (Clinical Genetics Society, 1994) emphasised the great diversity of attitudes and practices both within and between various professional groups. In general terms, it was apparent that genetic nurses and co-workers were more 'protective' of children than were the clinical geneticists, who were themselves more 'protective' than paediatricians or other medical specialists. The geneticists were more selective in the tests that they did or would perform on children, whereas the paediatricians expressed their willingness to comply with parental requests in most settings. The interpretation of these findings by the Working Party was that most paediatricians had not been faced by these issues very much in practice, while the geneticists had had more opportunity to reflect upon difficult cases and adverse outcomes of a range of counselling scenarios.

Although many professionals and other groups recommend caution in relation to genetic testing in childhood, there is little systematic evidence of it causing any harm. This is scarcely surprising because such testing has only recently become possible for most conditions, and there has been an understandable reluctance to offer predictive tests to children unless there is some medical benefit to be gained - this reluctance perhaps being one manifestation of 'appropriate paternalism'. The absence of evidence that such testing causes harm is not a reason for going ahead with such testing, particularly when there are good grounds for caution. Not only is there the abrogation of the child's rights to autonomy and confidentiality to consider, and the disadvantages of discrimination and stigmatisation that face adults with unfavourable genetic test results in relation to both predictive tests and tests of carrier status (Billings et al., 1992), but there is also the question of the impact of test results on the emotional and social development of the child. Given this lack of evidence, there has so far been a considerable degree of consensus among published reports and guidelines in relation to the predictive testing of children for late-onset disorders, recommending great caution (Clinical Genetics Society, 1994; Wertz et al., 1994; American Society of Human Genetics/American College of Medical Genetics, 1995) British Paediatric organisation of lay support groups in Britain - responded critically to the Clinical Genetics Society report, but they opposed such testing even more firmly, although adopting a more liberal approach to carrier testing (Dalby, 1995).

The question of evidence, or the lack of it, is primarily relevant in relation to the social and emotional consequences of testing; it has much less to say about the abrogation of rights, and to what extent that is a matter for concern. What is required now is the careful study of family and professional strategies for transmitting genetic information, including assessments of the success of various strategies in different family circumstances. It would be very unhelpful at this stage of our understanding simply to carry out widespread genetic testing of children and to wait the many years necessary for evidence of harm to accumulate. It would not be feasible to mount a randomised, controlled trial of two different approaches because there are so many dimensions to this issue that it could not meaningfully be reduced to a simple dichotomy. It will be most helpful now to record current practice systematically, and to base advice to families in a broad range of different contexts on the analysis of this experience.

While 'evidence' is important and helpful both in arriving at general policy and in making decisions about individual cases, it would be unwise to ignore the equally important roles of ethical reflection and conceptual clarification. Calls for more evidence and for further empirical studies do chime with the times in this era of evidence-based medicine, but it is all too frequently forgotten that 'facts' do not exist independently of the observer but are constructed socially by individuals and groups who have explicit opinions, implicit attitudes and vested interests which will shape these 'facts'. The pose of being a neutral, objective observer of one's fellow humans, able to discover 'the truth' about some aspect of mankind, has a long history but has been intellectually untenable for the best part of a century.

A second area in which research will be helpful, as pointed out by Michie and Marteau (1996) as well as Alderson (1993), is the maturation of the child's competence to participate in decision-making. This may help us to make more appropriate judgements in that grey area where the 'child' is able to make some contribution to important decisions about genetic testing. Finally, I will add that there has been considerable debate among professionals about childhood genetic testing over the past few years

(Clarke and Flinter, 1996; Michie, 1996; Michie et al., 1996; Clarke, 1997,; Clarke and Harper, 1992; Harper and Clarke, 1993; Chapple et al., 1996). That the issue has aroused such interest is a very positive development, because it used to be neglected - it was an invisible problem. On the other hand, it is disappointing that some parties have tended to polarise the debate in an oversimplistic fashion, portraying one camp as opposed to testing and another as for it. Let us hope that the next few years sees the debate widen to include other standpoints, and that professionals and families can move forward together towards a common approach to the management of these difficult issues.

REFERENCES

Alderson, P., 1993. *Children's consent to surgery*, Open University Press, Buckingham, England and Pennsylvania, USA.

American Society of Human Genetics/American College of Medical Genetics, 1995. 'Report. Points to consider, ethical, legal and psychological implications of genetic testing in children and adolescents', *American Journal of Human Genetics*, 57, pp. 1233-1241.

Bachman, R.P. et al., 1993. 'Compliance with childhood cholesterol screening among members of a prepaid health plan', *American Journal of Diseases of Children*, 147, pp. 382-385.

Ball, D. et al., 1994. 'Predictive testing in adults and children', in A. Clarke (ed), *Genetic Counselling, Practice and Principles*, Routledge, London, pp. 63-94.

Bergman, A.B. and Stamm, S.J., 1967. 'The morbidity of cardiac nondisease in schoolchildren', *New England Journal of Medicine*, 276, pp. 1008-1013.

Billings, P.R., 1992. 'Discrimination as a consequence of genetic testing', *American Journal of Human Genetics*, 30, pp. 476-482.

Binedell, J. et al., 1996. 'Huntington's disease predictive testing, the case for an assessment approach to requests from adolescents', *Journal of Medical Genetics*, 33, pp. 912-918.

Bloch, M. and Hayden, M.R., 1990. 'Opinion, Predictive testing for Huntington disease in childhood, challenges and implications', *American Journal of Human Genetics*, 46, pp. 1-4.

Bloch, M., 1992. 'Predictive testing for Huntington disease in Canada, the experience of those receiving an increased risk, *American Journal of Medical Genetics*, 42, pp. 499-507.

British Paediatric Association Ethics Advisory Committee, 1996. *Testing children for late onset of genetic disorders*, British Paediatric Association, London.

Buckman, R., 1992. *How To Break Bad News, a guide for health-care professionals*, Papermac, London.

Chapple, A. et al., 1996. 'Predictive and carrier testing of children, professional dilemmas for clinical geneticists', *European Journal of Genetics in Society*, 2, pp. 28-38.

Clarke, A., 1997. 'Parents' responses to predictive genetic testing in their children', *Journal of Medical Genetics*, 34, p. 174.

Clarke, A. and Harper, P.S., 1992. 'Genetic testing for hypertrophic cardiomyopathy', *New England Journal of Medicine*, 327, p. 1175.

Clarke, A. and Flinter, F., 1996. 'The genetic testing of children, a clinical perspective', in T.M. Marteau and M.P.M. Richards (eds), *The Troubled Helix: social and psychological implications of the new human genetics*, Cambridge University Press, Cambridge, England, pp. 164-176.

Clayton, E.W., 1995. 'Removing the shadow of the law from the debate about genetic testing of children', *American Journal of Medical Genetics*, 57, pp. 630-634.

Clinical Genetics Society, 1994. 'Report of the Working Party on the Genetic Testing of Children' (A.Clarke, chairman), *Journal of Medical Genetics*, 31, pp. 785-797.

Codori, A-M. and Brandt, J., 1994. 'Psychological costs and benefits of predictive testing for Huntington's disease', *American Journal of Medical Genetics* (Neuropsychiatric Genetics), 54, pp. 174-184.

Craufurd, D. and Harris, R., 1986. 'Ethics of predictive testing for Huntington's chorea, the need for more information', *British Medical Journal*, 293, pp. 249-251.

Craufurd, D. et al, 1989. 'Uptake of presymptomatic testing for Huntington's disease', *Lancet*, ii, pp. 603-5.

Dalby, S., 1995. Genetics Interest Group response to the UK Clinical Genetics Society report 'The genetic testing of children', *Journal of Medical Genetics*, 32, pp. 490-491.

European Community Huntington's Disease Collaborative Study Group, 1993. 'Ethical and social issues in presymptomatic testing for Huntington's disease, a European Community collaborative study', *Journal of Medical Genetics*, 30, pp. 1028-1035.

Fanos, J.H. and Johnson, J.P., 1995a,. 'Perception of carrier status by cystic fibrosis siblings', *American Journal of Human Genetics*, 57, pp. 431-438.

Fanos, J.H. and Johnson, J.P., 1995b. 'Barriers to carrier testing for adult cystic fibrosis sibs, the importance of not knowing', *American Journal of Medical Genetics*, 59, pp. 85-91.

Harper, P.S. and Clarke, A., 1990. 'Should we test children for 'adult' genetic diseases?' *Lancet*, 335, pp. 1205-6.

Harper, P.S. and Clarke, A., 1993. 'Screening for hypertrophic cardiomyopathy', *British Medical Journal*, 306, pp. 859-860.

Hoffmann, D.E. and Wulfsberg, E.A., 1995. 'Testing children for genetic predispositions, is it in their interest?', *Journal of Law, Medicine and Ethics*, 23, pp. 331-344.

Huggins, M. et al., 1992. 'Predictive testing for Huntington Disease in Canada, Adverse effects and unexpected results in those receiving a decreased risk', *American Journal of Medical Genetics*, 42, pp. 508-515.

Kessler, S., 1988. 'Invited essay on the psychological aspects of genetic counselling. V. Preselection, a family coping strategy in Huntington Disease', *American Journal of Medical Genetics*, 31, pp. 617-621.

Lawson, K. et al., 1996. 'Adverse psychological events occurring in the first year after predictive testing for Huntington's disease', *Journal of Medical Genetics*, 33, pp. 856-862.

Lifshitz, F. and Moses, N., 1989. 'Growth failure. A complication of dietary treatment of hypercholesterolemia', *American Journal of Diseases of Children*, 143, pp. 537-542.

Michie, S., 1996. 'Predictive genetic testing in children, paternalism or empiricism ?' in T.M. Marteau and M.P.M. Richards (eds), *The Troubled Helix, social and psychological implications of the new human genetics*, Cambridge University Press, Cambridge, England, pp. 177-183.

Michie, S. and Marteau, T.M., 1996. 'Predictive genetic testing in children, the need for psychological research', *British Journal of Health Psychology,* 1, pp. 3-16.

Michie, S. et al., 1996. 'Parents' responses to predictive genetic testing in their children, report of a single case study', *Journal of Medical Genetics,* 33, pp. 313-318.

Montgomery, J., 1993. 'Consent to health care for children', *Journal of Child Law,* 5, pp. 117-124.

Montgomery, J., 1994. 'Rights and interests of children and those with mental handicap', in A. Clarke (ed), *Genetic Counselling, Principles and Practice,* Routledge, London, pp. 208-222.

Morris, M. et al., 1988. 'Adoption and genetic prediction for Huntington's disease', *Lancet,* ii, pp. 1069-70.

Morris, M. et al., 1989. 'Problems in genetic prediction for Huntington's disease', *Lancet,* ii, pp. 601-3.

Newman, R.T. et al., 1990. 'The case against childhood cholesterol screening', *Journal of the American Medical Association,* 264, pp. 3003-3005.

Parsons, E.P. and Atkinson, P., 1992. 'Lay constructions of genetic risk, *Sociology of Health and Illness,* 14, pp. 437-455.

Parsons, E.P. and Atkinson, P., 1993. 'Genetic risk and reproduction, *Sociology Review,* 41, pp. 679-706.

Parsons, E.P. and Clarke, A., 1993. 'Genetic risk, women's understanding of carrier risk in Duchenne muscular dystrophy', *Journal of Medical Genetics,* 30, pp. 562-566.

Pelias M.Z., 1991. 'Duty to disclose in medical genetics, a legal perspective', *American Journal of Medical Genetics,* 39, pp. 347-354.

Sharpe, N.F., 1993. 'Presymptomatic testing for Huntington Disease, is there a duty to test those under the age of eighteen', *American Journal of Medical Genetics,* 46, pp. 250-253.

Thelin, T. et al., 1985. 'Psychological consequences of neo-natal screening for alpha-1-antitrypsin deficiency', *Acta Paediatrica Scandinavica,* 74, pp. 787-793.

Tibben, A. et al., 1992. 'DNA-testing for Huntington's disease in The Netherlands, a retrospective study on psychosocial effects. *American Journal of Medical Genetics,* 44, pp. 94-99.

Tyler, A. et al., 1992. 'Presymptomatic testing for Huntington's Disease in Wales 1987-1990', *British Journal of Psychiatry,* 161, pp. 481-489.

Wertz, D.C. et al., 1994. 'Genetic testing for children and adolescents, who decides?', *Journal of the American Medical Association,* 272, pp. 875-881.

World Federation of Neurology, 1989. 'Research Committee Research Group Ethical issues policy statement on Huntington's disease molecular genetics predictive test', *Journal of Neurological Sciences,* 94, pp. 327-332 and *Journal of Medical Genetics,* 1990. 27, pp. 34-38.

Index

C

D